D0142027

This book introduces the reader to statistical reasoning and its use in physics. It is based on a course developed for non-science majors at Cornell, and differs from other treatments by its wide-ranging use of quantitative methods, which are built up in a constructive way and assume only that the reader can add, subtract, multiply, and divide with confidence.

The author begins with a self-contained introduction to the everyday uses of probability, including the quantitative assessment of statistical information. Following a chapter on useful mathematical concepts, he develops the basic notions underlying mechanical motion and the molecular theory of gases. These ideas are then brought together with the theory of probability to introduce entropy as a measure of molecular agitation, and to develop an understanding of thermal equilibrium and the limitations on the conversion of heat to useful work. The final three chapters deal with the physics of the direction of time, chaos (the extreme sensitivity of many mechanical motions to initial conditions), and the role of probability in quantum mechanics. To aid self-instruction, there are solved problems at the end of each chapter.

The main application for this volume will be as a text for non-science students. However, the originality of the ideas and approach will also make this a valuable book for a public ranging from physics undergraduates to general readers.

Reasoning About Luck:
probability and its uses in physics

Reasoning About Luck:

probability and its uses in physics

VINAY AMBEGAOKAR

Department of Physics Cornell University Ithaca, New York

CAMBRIDGE
UNIVERSITY PRESS

Published by the Press Syndicate of the University of Cambridge
The Pitt Building, Trumpington Street, Cambridge CB2 1RP
40 West 20th Street, New York, NY 10011-4211, USA
10 Stamford Road, Oakleigh, Melbourne 3166, Australia

First published 1996

Printed in Great Britain at
the University Press, Cambridge

A catalogue record of this book is available from the British Library

Library of Congress cataloguing in publication data

Ambegaokar, Vinay.
   Reasoning about luck: probability and its uses in physics / Vinay Ambegaokar.
     p. cm.
   ISBN 0-521-44217-6 (hc). – ISBN 0-521-44737-2 (pbk.)
   1. Statistical physics. 2. Probability measures. I. Title.
QC174.8.A43 1996
530.1′592–dc20 95-47481 CIP

ISBN 0 521 44217 6 hardback
ISBN 0 521 44737 2 paperback

TAG

For my daughters

Nor but in merriment begin a chase,
Nor but in merriment a quarrel...

W. B. Yeats

Contents

Preface

This book has grown out of a course I have taught five times during
the last 15 years at Cornell University. The College of Arts & Sciences
at Cornell has a 'distribution requirement in science,' which can be
fulfilled in a variety of ways. The Physics Department has for many
years offered a series of 'general education' courses; any two of them
satisfy the science requirement. The descriptions of these courses in
the Cornell catalog begin with the words: 'Intended for non-scientists;
does not serve as a prerequisite for further science courses. Assumes no
scientific background but will use high school algebra.' This tradition
was begun in the 1950s by two distinguished physicists, Robert R.
Wilson and Philip Morrison, with a two-semester sequence 'Aspects
of the Physical World,' which became known locally as 'Physics for
Poets.' At the present time some three or four one-semester courses
for non-scientists, 'Reasoning about Luck' sometimes among them,
are offered each year.

What I try to do in this book and why is said in Chapter 1, but
some words may be useful here. I started the enterprise lightheartedly
hoping to do my bit to combat the widely perceived problems of
scientific illiteracy and – to use a fashionable word – innumeracy, by
teaching how to reason *quantitatively* about the uses of probability
in descriptions of the natural world. I quickly discovered that the
italicized word makes for great difficulties. Quantitative reasoning
in physics requires more mathematics than is learned in high school
algebra.

Nonetheless, my course as it has developed does abide by the
rules summarized in the descriptive sentences quoted in the opening
paragraph. Assuming at the start only that the reader can add,
subtract, multiply, and divide with confidence, I attempt to build up a
certain amount of useable mathematics in a constructive way. Given

these tools, I found that I could strike off more boldly than I had originally planned. To many teachers, and particularly to those with experience in teaching elementary physics, the subject matter is likely to seem eccentric – if not mad – as regards both the physics and the mathematical skills assumed to be teachable. Therefore it is important that I immediately share my experiences as a teacher of the contents of this book with a prospective reader or a prospective user of it as a text. To the former I have to say that it is not light reading. Only by working through it with a pencil and paper and understanding the solved problems at the ends of the chapters are you likely to get very far. Let me add, before I frighten away every possible purchaser, that a scientifically inclined high school senior could, I believe, read the book from cover to cover and profit from it. For this reason I think that the book should be in high school libraries. The first four chapters, in particular, form a self-contained introduction to the everyday uses of probability, and will be accessible to high school seniors.

My primary target is university teachers, particularly those like myself who do not normally teach non-scientists and may have avoided it because of a concern that it would in their hands become a dispensing of watery baby-food. For them, let me describe the students to whom it has been a delight to teach a course based on this book. They have typically been free of any background in science, but also free of math-anxiety. They have also been in some way mature enough to get into the spirit of the thing, which is to learn about a way of thinking that might otherwise remain foreign and about a subject that might otherwise be a closed door. One such student, a Theater-Arts major as I recall, told me that she never thought she would be grateful to Cornell for having a science requirement; another, an English major, said that he would not have guessed in advance that the matters discussed were accessible to him; yet another, a student of Spanish Literature, did so well on a question involving a heat engine in the final examination that I asked him where he had learnt the subject, to which he replied somewhat indignantly, 'In this course!' Compensating for such pleasurable experiences have been students who would have been better served by a more descriptive or a more conventionally structured introduction to physics. Originally I accepted every non-scientist who registered for the course; I now pass out a diagnostic test in high school algebra and suggest alternatives to those who have difficulty with it. Others who have found the course confusing have had enough preparation to enroll in a two-semester introduction to physics. I tell such students, if I can identify them early, that a more systematic approach might suit them better.

In its later manifestations the course also had an experimental component. The students did five experiments in an auto-tutorial laboratory. These experiments were done using equipment intended for another course, and depended on the goodwill of colleagues. Written instructions on how to perform the measurements were largely borrowed from the other course and somewhat impromptu. They are not included in this book. I mention the experiments because I believe they had educational value and were well received.

Many people have helped me in this enterprise. The writing was begun, more years ago than I care to admit, during the tenure of a John Simon Guggenheim Memorial Fellowship at the University of California in Los Angeles. It has been worked on at Cornell, at NORDITA in Copenhagen, and at the Åbo Akademi in Finland. At various stages, draft chapters have been read and encouragement offered by many friends and colleagues: to list them all would be idle name-dropping, but I do thank each and every one of them. Four Cornell graduate students, Boldizsár Jankó, Eric Grannan, Tracy Noble, and Robert Smith helped me in various ways; the last, in particular, devised some of the problems. Alexander Fetter, Benjamin Widom, and Joel Lebowitz annotated an early draft of a part of the book; I have taken many of their suggestions. Michael Fisher and David Mermin read a late version in great detail, covering it with invaluable marginalia, and David continued to read and advise as I rewrote. Louis Hand and Simon Capelin advised me on the final draft. I am truly grateful for all this help, but must and do accept responsibility for the errors and idiosyncrasies that remain.

Vinay Ambegaokar   *Ithaca, New York*

1

Introduction

The eternal mystery of the world is its comprehensibility
Albert Einstein

The purpose of this little book is to introduce the interested non-scientist to statistical reasoning and its use in physics. I have in mind someone who knows little mathematics and little or no physics. My wider aim is to capture something of the nature of the scientific enterprise as it is carried out by physicists – particularly theoretical physicists.

Every physicist is familiar with the amiable party conversation that ensues when someone – whose high school experience of physics left a residue of dread and despair – says brightly : 'How interesting! What *kind* of physics do you do?' How natural to hope that passing from the general to the particular might dispel the dread and alleviate the despair. Inevitably, though, such a conversation is burdened by a sense of futility: because there are few common premises, there is no reasonable starting point. Yet it would be foolishly arrogant not to recognize the seriousness behind the question. As culprit or as savior, science is perceived as *the* force in modern society, and scientific illiteracy is out of fashion.

However much I would like to be a guru in a new surge toward literacy in physics, ministering to the masses on television and becoming rich beyond the dreams of avarice, this, alas, is not to be. Among other things, I am immune to descriptions of science which, even when transported by the enthusiasms and exaltations of the teller, are indistinguishable from fairy stories because they offer the reader no way of questioning or reasoning about what is being told. It is precisely the questioning and reasoning listener that physics addresses.

A parable may illustrate the point. Some years ago at Cornell, Hans Bethe, whose discoveries and leadership had made him a magisterial figure in physics, was interrupted in mid-equation by a young man who said: 'No. That's not right.' Pause. The interruption was only a shade more categorical than many that occur during seminars in theoretical physics. Fingers against jaw, Bethe pondered for a few more moments. Then, in a Germanically accented American that his acquaintances will be able to hear he said: 'Gosch …back to the drawing board.' A small part of his reasoning had been demolished. An exchange like this is possible because mathematics is the language of theoretical physics, and reasonable people can agree when a small mathematical error has been made. Could something like this happen at a lecture by a famous historian or deconstructionist critic? Not, I think, in the same way.

To engage in a discourse with you, the reader, in which I try to introduce you to new ideas and to offer you the wherewithal to say: 'No. That's not right,' I see no way out of using elementary mathematics. If those are alarming words, I hasten to add, soothingly, that you already know much of what is needed and that this book will teach you the rest.

A few paragraphs back I used the words literacy in physics. What does the expression mean to me; and in what sense will reading these rather dithyrambic pages confer it on you? Well, there are many levels of literacy. To read some of the poems of T.S. Eliot one would seem to need some acquaintance with the ancients in Greek and Latin, a smattering of Dante, a knowledge of French Symbolist poetry, and bits and pieces of the Upanishads and the Bhagvat Gita – a German translation of the latter being permitted. All this may be needed to comprehend Eliot thoroughly. But, given that life is too short for thoroughness in all matters, it is better to have a deeply felt and personal sense of a few poems than to slog superficially through many.

Physics is a body of knowledge and a point of view. Both have grown out of the astonishing discovery in seventeenth century Europe that the observed workings of inanimate nature, from how the planets move to how a prism makes a rainbow, can be accurately summarized in mathematical terms. Why the world is thus regulated and why we have evolved to a point a little beyond cats from which we can perceive these regularities are questions for which science has no answers.†

That mathematical relationships ('Laws') do exist is demonstrably true. The physical laws that have been discovered – by a mixture of observation, intuition, and a desire for a concise and therefore in some sense beautiful description – do not merely organize experience, they organize it in a manner that encourages disprovable predictions. The wave theory of light predicted a bright spot in the center of

† If you are interested in reading about attempts at answers, the key-words *Anthropic Principle* will get you started, at your own risk, in a good library.

the shadow cast by a circular screen, a result so apparently absurd that it made believers of disbelievers when it was found. In physics, the search for unifying laws is a continuing quest. Great unifications have, however, occurred rarely; when they do they become part of the physicist's general knowledge and confer well-deserved immortality on their discoverers. The daily work of most theoretical physicists is more mundane. Some, working in areas where the laws are not well-formulated, have the satisfaction of constructing bits and pieces of mathematical structure that explain on general grounds why observed phenomena occur and unobserved do not, building blocks for greater constructions to come. Others are able to understand how an apparently bizarre observation is a consequence of accepted theory, or – and this is a great joy – of predicting quantitative relations between measurements, or phenomena few would have anticipated. When there are disagreements it is often unclear whether the observations, the mathematics, or both are at fault. The resolution of conflicts is emotional and often heated. More than in most fields of scholarship, though, when the dust settles, it settles for good. This suggests that the physical study of nature is an act of uncovering a hidden structure. That we are inextricably part of the structure makes its study all the more fascinating.

In this description of physics as the work done by physicists I have passed from the seventeenth century to our times with lightning speed. In scientific matters there *is* an unusual kinship through the ages. Consider some modern technology and ask what branches of physics come into play and what names emerge as the earliest contributors whose work is still actively in use. In this game I shall ignore, without in any way intending to belittle, the scientific and engineering genius that connects the basic physics with the manufactured product. What shall we pick? Space is in the news these days. The design of a modern space vehicle requires the following interrelated disciplines: the mechanics, or science of motion, of Isaac Newton (1642–1727); the thermodynamics, or science of heat, anticipated by Sadi Carnot (1796–1832) who called it 'La puissance motrice du feu' – the motive power of fire – and formulated by Rudolf Clausius (1822–88) and Lord Kelvin (1824–1907); and, the hydrodynamics, or theory of the flow of fluids, for which the first important steps were taken by Daniel Bernoulli (1700–82) and Leonhard Euler (1707–83). The sophisticated electronic and communication systems are ultimately based on: the unified theory of electricity, magnetism, and light of James Clerk Maxwell (1831–79); the statistical mechanics associated with Ludwig Boltzmann (1844–1906) and Josiah Willard Gibbs (1839–1903); and

the quantum mechanics invented by Werner Heisenberg (1901–76) and
Erwin Schrödinger (1887–1961). In fact, the subjects listed, together
with the special theory of relativity of Albert Einstein (1879–1955),
who modified Newton's mechanics to unify it with Maxwell's elec-
tromagnetism, taking into account certain unexpected properties of
light, make up an introduction to the study of physics, or a base for
the study of engineering. Someone who has worked his or her way
through this curriculum is, unquestionably, literate in physics.

A two-year course of study with mathematical prerequisites, needed
to complete the above program, is too much to ask of someone with
no professional interest in science who, nonetheless, would like some
insight into what physics is about. It is possible to conduct a survey
of Great Ideas in Physics, colloquially called 'Physics for Poets.' Such
a survey is not at all easy to do and would at my hands become a
once-over-too-lightly affair: the brothers Grimm instead of Galileo.†
So, when my chance came to teach a course on science for non-
scientists, I decided to try something different. For the purpose at
hand, I saw no particular virtue in completeness and no particular
vice in the unconventional. My course, which had the same title
and subject matter as this book, attempted to convey the way in
which physicists think about irreversibility and entropy, heat and
work, and, very briefly, quantum mechanics: conceptually interesting
topics tied together by the need for probabilistic concepts. An intuitive
introduction to the mathematics of probability was provided. I had
not anticipated that among the students who later seemed to profit
from the course would be some whose recollection of high school
algebra was hazy. Some mathematical folk-remedies were, therefore,
also thrown in. Here is the description written for the second offering
of the course:

'A course for inquiring non-scientists and non-mathematicians which
will attempt to explain when and how natural scientists can cope rationally with chance. Starting from simple questions – such as how
one decides if an event is 'likely,' 'unlikely,' or just incomprehensible
– the course will attempt to reach an understanding of more subtle
points: why it is, for example, that in large systems likely events can
become overwhelmingly likely. From these last considerations it may
be possible to introduce the interested student to the second law of
thermodynamics, that putative bridge between C.P. Snow's two cul-

---

† Galileo Galilei (1564–1642), arguably the first modern physicist, who said, about the 'book' of
Nature: '…it cannot be read until we have learned the language and become familiar with the
characters in which it is written. It is written in mathematical language.'

tures. Another physical theory, quantum mechanics, in which chance occurs, though in a somewhat mysterious way, may be touched on.

'The course is intended for students with not much more preparation than high school algebra. The instructor will from time to time use a programmable pocket calculator to do calculations with class participation. Some of the key ideas will be introduced in this way.

'The dearth of appropriate readings for a course of this kind was a source of complaint during its first trial. As of this writing, the instructor has high hopes of producing lecture notes or a very rough draft of a short book. Here are some questions that a student may expect to learn to answer. In a class of 26 people why is there (and what does it mean that there is) a 60% chance that two persons have the same birthday? If 51% of 1000 randomly selected individuals prefer large cars to small, what information is gleaned about the car preferences of the population at large? Why does a cube of ice in a glass of soda in your living room never grow in size? – a silly question that can be answered seriously. What can and cannot be said about the way in which the image appears on a developing photographic plate?'

As it turned out, I was able to get somewhat further than I anticipated, but I never did write the short book that was promised. Here it is. I have found it very hard to do, and have often wished that I had not so irrevocably promised to explain everything honestly. But I did so promise, and I think I have so done. You will decide if it was worth it.

A word of warning about the virtues and vices of this book. For every topic covered in some detail, one nearby which any systematic treatment would include is ignored. Perhaps Julia Child says it best in *Mastering the Art of French Cooking*, 'No pressed duck or *sauce rouennaise*? No room!'

2

The likely, the unlikely, and the incomprehensible

Lest men suspect your tale untrue
Keep probability in view

John Gay

The mathematical theory of probability was born somewhat disreputably in the study of gambling. It quickly matured into a practical way of dealing with uncertainties and as a branch of pure mathematics. When it was about 200 years old, the concept was introduced into physics as a way of dealing with the chaotic microscopic motions that constitute heat. In our century probability has found its way into the foundations of quantum mechanics, the physical theory of the atomic and subatomic world. The improbable yet true tale of how a way of thinking especially suited to the gambling salon became necessary for understanding the inner workings of nature is the topic of this book.

The next three chapters contain some of the basic ideas of the mathematical theory of probability, presented by way of a few examples. Although common sense will help us to get started and avoid irrelevancies, we shall find that a little mathematical analysis yields simple, useful, easy to remember, and quite unobvious results. The necessary mathematics will be picked up as we go along.

In the couplet by John Gay (1688–1732), the author of the Beggar's Opera, probability has a traditional meaning, implying uncertainty but reasonable likelihood. At roughly the same time that the verse was written, the word was acquiring its mathematical meaning. This first occurs in English, according to the Oxford English Dictionary, in the title of a book, published in 1718, by Abraham de Moivre (1667–1754), an English mathematician of French Hugenot extraction: *The Doctrine of Chances: A Method of Calculating the Probability of Events in Play.* The analysis of games of chance had its correct modern beginnings in France in the 1650s and attracted the attention of thinkers like Blaise Pascal (1623–62), Pierre de Fermat (1601–65), and Christiaan

Huygens (1629–95) – all of whom also made important contributions to physics.†

Everyday words in the scientific vocabulary – field, charge, and strangeness are examples – usually have precise technical meanings remote from their ordinary ones and much less well known. 'Probability' was taken over so long ago and used so aptly and the concept applies or is thought to apply to so many situations that the technical meaning has slowly edged its way onto center stage. In mathematics and in science generally a numerical value is attached to the word: uncertain events are rated on a scale from zero to one; something that is unlikely is said to have a low probability. Although this is now part of the common general vocabulary, it deserves a more precise statement.

The technical use of the word probability applies in general to a special class of situations. It presupposes a repeatable experiment or observation with more than one possible outcome controlled by chance, which means that before the fact, precisely which outcome will occur is neither known nor deducible. For such an experiment, the probability of a given outcome is a numerical estimate, based on experience or theory, of the fractional occurrence of that outcome in a large number of trials.‡

Several questions are raised by this definition. What observations are both repeatable and uncertain? What are the possible outcomes? How are probabilities assigned? What precisely is the meaning of a 'large number' of trials? What good is a concept that is, on the face of it, so uncertain?

Since our subject was born at the gaming table, it is not surprising that these questions are most easily answered in the context of gambling. They are harder to address in less controlled situations where the notion of probability may nonetheless be useful, except by hypothesizing unprovable analogies with games of chance.

Let us start then by considering the rolling of an ordinary six-sided die. It is clear that this is something that can be done over and over again, and that the outcome in each case will be that one or another of the faces is uppermost. (It is natural to ignore as misthrows rolls with other outcomes, e.g. becoming wedged at an angle in the pile of a carpet.) Unless the die is launched by a very precise machine or a *very* clever cheat, there is enough variability in the experiment to rule out

† Less than correct beginnings are to be found in the writings of Gerolamo Cardano (1501–76).
‡ A probability equal to unity thus means that a particular outcome always occurs. Saying that something has a probability of one is a complicated way of describing a certainty as a limiting case of probability.

the possibility of predicting the result of any given throw. The roll of a die is, with the minor caveats thrown in to calm excessively logical minds, a clear cut example of a repeatable experiment with random outcomes.

Less straightforward examples emerge from a listing of a few areas in which statistical reasoning, i.e. reasoning based on probabilities, is useful. These range from (a) quality control – a machine producing ball-bearings nominally 5 millimeters in diameter has a probability of producing an oversize one with a diameter greater than 5.03 millimeters – to (b) marketing – an individual in a certain city has a probability of preferring brown eggs to white, when both are of the same size and price; from (c) epidemiology – someone who has been vaccinated against a disease has a probability (smaller, one would hope, than someone who has not) of contracting the ailment, during a given summer in a particular section of the country – to (d) genetics – the flowers produced by a plant of given ancestry have a probability of being blue, as opposed to white or mixed. Both experimental and theoretical physics provide unusually nice examples. Experimental data contain random errors, amenable to statistical analysis. The statistical structures in theoretical physics will occupy this book from Chapter 7 on.

Each of the above examples contains a random event: (a) the production of a ball-bearing by a given machine; (b) the choice by an individual between brown eggs and white of the same size and price; (c) the moving about in a specified environment of a vaccinated person who then does or does not contract the disease in question; (d) the flowering of a plant produced by crossing given other plants. The role of chance is also evident. Even when a reasonable attempt is made to avoid irrelevant comparisons, such as comparing the output of two quite different machines, the repeated observations are only superficially identical. The gears and shafts of the machine are in different positions as different ball-bearings are produced, the grinding surfaces are wearing, there are varying levels of vibration in the moving parts, and so on. In fact, until we enter the unusual world of quantum mechanics in Chapter 12 probabilities will occur, as in games of chance, only as a consequence of imprecise specifications or missing information.

It is also clear in each of the cases discussed above that the outcomes can be divided into mutually exclusive groups such that any event falls into one or another of the groups.

These examples are all more straightforward than some common situations to which the concept of probability is applied. When the TV weatherperson says that the probability of rain tomorrow is 20%, what do we understand? We do not, of course, pedantically insist that there is only one tomorrow, in the course of which it will either rain or not rain. Instead, we have to make an analogy with an event in a lottery or a game of chance – drawing a number, being dealt a hand of cards – which has several possible outcomes, and can, in principle, be repeated many times. Without realizing the complexity of the thought, we might do something like the following: imagine a large number of days during which the weather pattern is, in all ways that are thought to be significant, the same as it is today, and understand that the best estimate of the forecaster is that rain will fall on two out of ten of the days following these many, to all intents and purposes identical, days.

There are situations in which the repeatable experiment is even more hypothetical. When the theory of probability is used to design a telephone exchange, only one with a large probability – on the basis of some model of peak usage – of not being overloaded with calls is actually constructed.

Repeatability, at least in principle, is crucial. If I ask you the probability of your going home this Thanksgiving, and we are speaking about you uniquely, the word is being used in a literary sense, and the question has no numerical answer. On the other hand, if we are speaking about you as one of a group, say a group that normally eats in a certain large dining hall of which I am the manager, the question is statistically meaningful, and the numerical value, which my experience would have taught me to estimate, is useful.

It is worth making the point that the concept of probability makes sense only when there is some understanding or working hypothesis about what is going on. This is well illustrated by a story told by my colleague Michael E. Fisher. When he was a student at King's College, London, a man seeking advice from someone skilled in statistics was directed to him. The inquirer's research task required that he enter in a ledger the results of urine analyses from a large number of London hospitals. He had begun to suspect, and was excited by the suspicion, that there were slightly more entries showing abnormally high sugar at the upper as opposed to the lower end of each page. Fisher asked the man if he had any hypotheses to explain the supposed effect. He did not. Did he start a new page every morning? Did the results come to him from different hospitals in some special order, or did he, himself, order the entries according to some plan or intuition? Were emergency tests treated differently from routine ones? Did the results come in batches of a given size? After many such questions, all of which were answered in the negative, Fisher concluded that it would be pointless to pursue the matter further, since there was not the remotest reason being offered for why reading the entries from top to bottom was in any way different from reading them in any other order. Even if the supposed connection between the top of the page and high sugar withstood statistical scrutiny, it would have to be treated as meaningless.

From the required minimum of a working hypothesis varying levels of theoretical understanding of an uncertain observation are possible, which brings us to the question of how probabilities are assigned. At one end of the scale, the probability can be calculable; at the other, the probabilities may be unknown numbers, to be estimated from accumulated experience. If an ordinary coin is tossed in an ordinary way, the probability of heads is $\frac{1}{2}$. Similarly, if an ordinary die is rolled in an ordinary way, the probability of any given face is $\frac{1}{6}$. The idea being used here is that when outcomes are equally likely the probability of any particular outcome is unity divided by the number of such outcomes. The question of what outcomes are equally likely is not without subtlety, and this 'principle' at best gives a working hypothesis for the probability to be assigned to an outcome. In particular, not all coins are fair. It should be possible from the shape, density, and manner of tossing to calculate the probability of heads for a loaded coin, but I have to admit that I wouldn't know how to do it. This number could, however, be estimated by tossing the coin many times and recording the fractional number of heads.

What is the connection between theoretical and empirical probabilities? Here is where the 'large number' of trials comes in. It is a remarkable fact, to be discussed in detail later, that the accumulation of more and more experience about a repeatable random experiment permits better and better estimates of the theoretical probabilities of the possible outcomes. A simple example may whet your appetite for what is to come. If a coin is tossed 1000 times and heads are obtained on 511 occasions, we shall find that this is quite compatible with the hypothesis that the coin is fair. On the other hand, if there are 511 000 heads in 1 000 000 tosses – the same percentage as in the smaller number of trials – there are extremely strong reasons for suspecting that the coin, the method of tossing, or both favor heads over tails in a ratio close to 51/49. The mathematical theory of this phenomenon, sometimes called the law of large numbers, also assigns quantitative probabilities to the vague words – 'extremely strong'; 'close to' – in the last sentence. At the risk of getting ahead of my story I will add that this little example, for which our everyday experience does not prepare us, contains the essence of why it is possible to reason quantitatively about luck.

Calculating probabilities

The Theory of Probability is concerned with reasoning from the probabilities of elementary events (of which a example might be drawing a particular card from a deck, or getting a six on one roll of a die) to those for compound events (such as being dealt a hand of cards of a particular sort, or getting a specified result in a throw of many dice). We shall not be going very deeply into this subject, which has remained an active branch of mathematics precisely because it is subtle and leads to unexpected results, but we shall find it rewarding to work out a few examples.

Here is a problem that goes back to the very beginnings of the subject. It has to do with the throwing of dice of the ordinary six-sided sort. A fair die is one for which all faces are equally likely, so that the probability of any particular face is, as we have already discussed, 1/6. If four fair dice are thrown, what is the probability of at least one six showing?

Problems of this kind require the calculation of, first, the number of possible outcomes, and, second, the number of desired outcomes. When, as in this case, the outcomes are equally likely, the probability, in complete analogy with the situation for a single die, is the second number divided by the first. Now, one die has six equally likely faces. For *each* of these faces, a second die has six equally likely outcomes. Thus there are 6 × 6, or 36, equally likely outcomes for two dice. A third die can have any of its six faces showing for each of the 36 possibilities for the first two, and so on. When four dice are thrown, there are thus 6 × 6 × 6 × 6 = 1296 equally likely outcomes.

How many of these are desired outcomes, containing at least one die with six dots uppermost? At first sight, this seems like a difficult question. 'At least one six' means all cases with one six, two sixes, three sixes, or four sixes. One can calculate the number of ways in which each of these classes of outcomes can occur, and we shall do such an ennumeration in the last section of this Chapter. However, it is possible to be more deft. Note that if we add the desired outcomes (call them successes) to all the other outcomes (call them failures) we must get the total number of outcomes that we have already calculated. But only the cases in which *no* six is showing are failures. How many of these are there? Well, there are five ways in which the first die can fail to have its six-face uppermost, and, for each of these, five ways of arranging the second die without having its six showing, and so on. So the number of failures is 5 × 5 × 5 × 5, or 625. The number of successful outcomes must therefore be 1296 − 625 = 671.

The desired probability is thus $671 \div 1296$, which equals $0.517\,746\,9\ldots$, a number slightly bigger than $\frac{1}{2}$. We see therefore that playing this game with four dice is entirely equivalent to tossing a loaded coin with the probability of one face being $671 \div 1296$ while that of the other is $625 \div 1296$. There is no intrinsic reason for the probability of an event to be a simple number; the only requirement is that it must be between zero and one. The sum of the probabilities for all mutually exclusive outcomes must be unity, since it is a ratio whose numerator and denominator are equal to the total number of outcomes.

Apart from enumerating all possible outcomes, which can very rapidly become very tedious and weary even a modern computer, there are no completely systematic ways of solving such problems. By learning how to do a few of them, one accumulates a store of useful tactical maneuvers and trains one's intuition to set off in the right direction. To show the cumulative effect of problem solving, and to have a little fun, let's try another example in which the trick we have just used is again useful.

What is the probability that in a group of N persons at least two have the same birthday? To start, we have to make some assumptions. First, we shall exclude, with apologies to those so slighted, individuals born on the 29th of February, because their inclusion makes the problem harder. Second, we shall assume that for the remaining people in a large population, from which we shall imagine selecting our N persons at random, every day is equally likely as a birthday. If it is a fact, unknown to me, that maternity wards have seasonal rushes and slack periods, then our calculation will be correspondingly, and surely at most very slightly, incorrect.

Enumerating all ways of having one or more coincidences of birth-dates is not easy, since this includes many possibilities. Here again it is simpler to calculate the probability of the complementary event, i.e. of the outcomes excluded in the question. First we can work out the denominator of the required ratio. In how many ways can birthdates occur among N individuals, without restrictions of any kind – except for the exclusion of February 29? There are 365 days in a normal year, and thus 365 possible birthdates for a given individual. A second person has the same number of possible birthdates for each possible one for the first. Thus there are 365×365 ways of assigning birthdays to two individuals. Similarly, a group of N persons has a total of $365 \times 365 \times \ldots 365$ possible assignments of dates of birth, where the dots signify that there are N factors of 365 multiplied together, a number that in the ordinary notation of algebra would be written as $(365)^N$.

Of this large number of possibilities – each of which, according to our assumption, is equally likely – how many correspond to situations in which no two individuals have the same birthday? Now once we have assigned a birthday to one individual, which, as before, can be done in 365 ways, there are only 364 possibilities for the second person. And so on. Our group, then, has $365 \times 364 \times \ldots (366 - N)$ possible assignments of *distinct* dates of birth. Here the dots indicate multiplication by numbers decreasing by unity, and the last one has been written algebraically in such a way as to express correctly the fact that there are a total of N factors in the expression. For example, if N were 3, the last factor would correctly come out to be 363, and, if N were 366, the last factor, and thus the product, would be zero as it should be: there is *no* way of assigning distinct days of an ordinary year to 366 people.

The ratio of the two numbers calculated above is the probability that N persons selected at random will have distinct birthdays. Every one of the cases not included in the second part of the calculation contains at least one coincidence of birthdates. Since we have divided all possibilities into two mutually exclusive categories, the sum of the cases in each category is equal to the total number of outcomes, and, as in the previous example, the sum of the probabilities for the two categories is one.

The answer to our problem is thus:

$$1 - \frac{365 \times 364 \times \cdots (366 - N)}{(365)^N}. \tag{2.1}$$

As it stands, such a formula says little. Its meaning emerges when special cases are worked out. Table 2.1 shows the results obtained from the general formula for various numerical values of N using a pocket calculator, and Fig. 2.1 is a graph of this relationship. Note that the probability increases quite rapidly with N, becomes larger than 0.5 for a group of only 23 people, is just under 0.6 for 26, and is all the way up to 0.97 for 50. Most people are surprised by these results, and without further thought they do violate one's intuition. It is important to recognize that we have *not* concluded that if you were to encounter 22 other individuals there would be a better than even chance that one of them would have *your* birthday; the probability of that occurring can be worked out and is a little less than 6%. The result obtained states that in a group of 23 there is a better than even chance that some unspecified person's birthday will be repeated. This is quite another matter, since there are many ways† of choosing a pair from such a group.

---

† 253 in fact – a number we shall soon learn to calculate.

Some rules

Now is a good time to pause and distill a few general principles. Three key questions have been asked in this chapter. When does the technical concept of probability apply? How are numerical values assigned to the probabilities for elementary events? How are the probabilities of complicated outcomes deduced from those for elementary ones? I have tried to answer the first two of these questions in a commonsense way, avoiding issues around which philosophical debates rage. The third question is more straightforward; it is answered by a set of mathematical rules, which could be presented as a logical axiomatic structure. Here again, I want to take a physicist's point of view, making sure that we know how to work out particular cases, and finding time for generalities only to the extent that they help in that endeavor.

We have, in fact, discovered three simple rules.

Rule (1): The probabilities for all mutually exclusive outcomes add up to unity.

In problems such as those done in the last section, the above addition yields a ratio whose numerator and denominator are each the total number of equally likely outcomes, and thus unity. More generally, note that this sum of probabilities is equal to the probability that an arbitrary one of the mutually exclusive outcomes will occur, i.e. it is the answer to a question like, 'What is the probability of heads *or* tails when a coin is tossed?' But that is a certainty, and a certainty has unit probability.

Table 2.1. *Probability p(N) of at least one coincidence of birthdates in groups of size N for various particular cases.*

Group size N	Probability p(N)
10	0.11695
20	0.41143
22	0.47570
23	0.50730
30	0.70632
40	0.89123
50	0.97037
60	0.99412
100	0.9999996

Rule (2): The probability for two *independent* events both to occur is the product of the probabilities for the separate events.

Recall our discussion of throwing several dice, where both the total number of outcomes and the number of 'unsuccessful' outcomes were obtained by multiplying the numbers for individual dice. The answer we obtained to the question, 'What is the probability of no six showing when four dice are rolled?' was $(5 \times 5 \times 5 \times 5)/(6 \times 6 \times 6 \times 6)$. But this is also $(\frac{5}{6} \times \frac{5}{6} \times \frac{5}{6} \times \frac{5}{6})$.

Rule (3): Even when two events are not independent, the probability of both occurring is the probability of the first event times the probability of the second, subject to the condition that the first has occurred.

An example of the third rule occurs in a calculation of the probability of two individuals having different birthdates. Looking back at our previous work, we see that this probability is $1 \times (364/365)$. The first factor is the probability of one person having any birthday

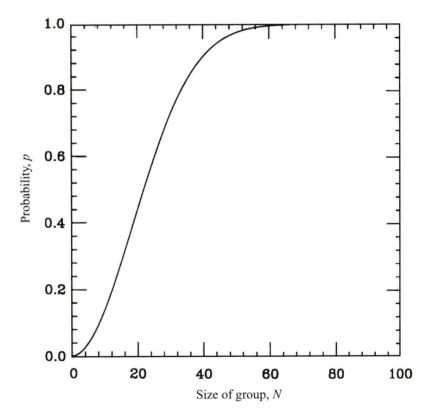

Fig. 2.1. Probability of at least one coincidence of birthdates vs. Size of group. [A smooth curve is shown only for convenience; since the size of the group is an integer, the plot should consist of a series of dots.]

at all, and the second is the probability of another having any *other* birthday. Similarly, the probability of three individuals having different birthdays was found to be (364/365) × (363/365). The first factor is the probability of two individuals having different birthdays, and the second is the probability of a third having any birthday *except* those of the first two.

There are other rules of this kind. For our purposes, it is less important to memorize them than to understand the thoughts that underlie them. Either way, with practice one quickly improves one's ability to answer questions of this kind, though I emphasize that it is no disgrace to be stumped by some that sound simple. This happens even to long-time professional scientists.

Here is a problem that came up in our discussion of birthdays. What is the probability that *none* of 22 randomly selected individuals has *your* birthday? The answer, using rule 2 is $(364/365)^{22}$, a number which my pocket calculator tells me is 0.941. So, using rule 1, the probability of at least one of them having your birthday is 0.059, corresponding to the slightly less than 6% that was mentioned earlier. There is no harm, and in fact some advantage, in reasoning this out step by step as we did in the last section, rather than simply using rules.

How to permute and combine

In counting possible outcomes, one frequently encounters a problem that can be phrased as the question: In how many ways can r objects be chosen from a collection of N objects? The answer depends, as we shall see, on whether one is or is not to worry about the order in which the r objects are chosen. The Oxford English Dictionary tells me that these problems were first referred to as Permutations (order distinguished) and Combinations (order not distinguished) in 1570 and 1673, respectively. They still are so named. Let us see how such questions would arise in a problem we have already met, and then work out the general situation for later use.

In the game with four dice, we managed to avoid explicitly counting the cases in which at least one six shows. Suppose that we had not been so clever. Then we would have had to enumerate all cases in which one, or two, or three, or four sixes showed. How is this to be done? Consider, first, the cases with one six. To make things concrete, we can imagine that the four dice have different colors, say red, blue, yellow, and green. Any one of these dice can have its six-face up. So we have to answer the trivial question: in how many ways can one chose this special die? The answer is, of course, in four ways. This part of the problem is related to the general problem posed in the last paragraph. In the language used there, the number of permutations of four objects taken one at a time is four. To finish the counting of cases in which four dice have exactly one six showing, we have to recognize that there are 5^3 or 125 ways of arranging the other faces so that no six is showing. Thus, the number of cases in the first category is 4×125 or 500. Now consider the cases with two sixes showing. In how many ways can one chose this preferred pair of dice? The first of them can be chosen in 4 ways, and the second in 3 ways, which is to say that there are 4×3 or 12 'permutations' of 4 objects taken 2 at a time. But this would be an incorrect way of counting: one does not want in this problem to distinguish between the pair red-blue and the pair blue-red. In short, it is the number of 'combinations' of four objects taken two at a time that is wanted, and this is evidently half as many as the number calculated, i.e, $12 \div 2$ or 6. This number has to be multiplied by $5^2 = 25$ to give the total number of cases with exactly two sixes showing, which gives a total of 150 such cases. Now we go one step further. Reasoning in exactly the same way as we did before, we find that there are $4 \times 3 \times 2$ 'permutations' of four objects taken three at a time, and $3 \times 2 \times 1$ ways of re-arranging three objects. In short, there are 24/6 or 4 'combinations' of four objects taken three at a time – try it imagining colored dice – and thus 4 ways of picking the three dice that show sixes, or $4 \times 5 = 20$ cases in which exactly 3 sixes are up. Finally, there is precisely 1 case in which all four dice show sixes. Thus our step by step enumeration has led to a total of $500 + 150 + 20 + 1 = 671$ cases, in satisfying agreement with the number we calculated rather more simply before.

We may systematize what we learned through special cases in the last paragraph by writing general expressions. Let $^N P_r$ denote the number of permutations of N things taken r at a time. Then, reasoning exactly as we did above we derive the formula

$$^N P_r = N \times (N-1) \times \ldots (N-r+1) \tag{2.2}$$

because there are N ways of chosing the first object, $(N-1)$ ways of chosing the second, and so on until we have multiplied together r factors. To calculate the number of combinations, which is conventionally written either as $^N C_r$ or as $\begin{pmatrix} N \\ r \end{pmatrix}$, we must divide $^N P_r$ by the number of ways in which r things can be re-ordered, or permuted, among themselves. This number is ^rP_r or $r \times (r-1) \times \ldots 1$. Thus we have

$$^NC_r = \binom{N}{r} = \frac{N \times (N-1) \times \ldots (N-r+1)}{r \times (r-1) \times \ldots 1}. \qquad (2.3)$$

The denominator on the right side of this equation occurs so often that it has its own name, $r!$ which is said 'r factorial.'

The above formula for the number of combinations of N things taken r at a time may be written in a more symmetrical form from which something we already discovered becomes self-evident. Looking back at the counting problem involving four dice, we recognize that in the language just introduced we had discovered that $\binom{4}{1} = \binom{4}{3}$, both these quantities being equal to 4. If on the right side of the last displayed equation (2.3) we multiply above and below by $(N-r)!$, what do we get? Notice that the numerator now becomes $N!$. So we find the neat formula

$$\binom{N}{r} = \frac{N!}{r! \times (N-r)!}. \qquad (2.4)$$

Note that the right hand side of this formula does not change when r is replaced by $N-r$. It thus follows that

$$\binom{N}{r} = \binom{N}{N-r} \qquad (2.5)$$

of which our discovery is the special case, $N = 4$, $r = 1$.

Terminology

Certain rather special usages conventional in the mathematical theory of probability have crept into this chapter. It may be useful to clarify some of these special meanings by attempting some definitions. The word 'experiment' has been used to describe something like the rolling of a die or dice. The word 'observation' has a very similar meaning but includes situations which are less controlled: we can 'experiment' with dice; we 'observe' the weather. A 'trial' is a repetition of an experiment or observation under reasonably identical conditions. An 'event' is an actual or realized outcome. [Although we have not had to use the words, the 'event-space' is the set of all possible distinct outcomes.] 'Mutually exclusive outcomes' refers to a division of the event-space into sectors such that each possible outcome belongs to only one sector.

These definitions are intended to clarify the basic concepts introduced in this chapter, which has also been concerned with showing how to deduce their consequences. The best way to see if one has made ideas one's own is to try them out on some concrete problems. I offer you some below.

Solved problems

(1) Two fair dice are rolled. What is the probability that at least one of them shows a three?
Answer: $1 - (5/6)^2 = 11/36$

(2) If a card is drawn at random from a standard deck of 52 cards – with no joker, what is the probability that it is a queen or a heart?
Answer: $(4 + 12)/52 = 4/13$

(3) Five fair coins are tossed. What is the probability that the number of heads exceeds the number of tails?
Solution: The answer is 0.5. Since there are an odd number of coins, there must either be more heads than tails, or vice versa. For every outcome in the first category, there is one in the second with the *same* probability, because it corresponds to reversing the roles of heads and tails. Thus, the number of 'successful' outcomes is exactly half the total number of outcomes. A perfectly correct but more laborious method would count the number of ways in which 0, 1, 2, 3, 4, or 5 heads could occur as 1, 5, 10, 10, 5, 1, adding up to the $2^5 = 32$ possible outcomes. [Note the 'symmetry' just mentioned: the probability of 0 heads is the same as the probability of 5 heads, because the latter is the probability of 0 tails, etc.] From this explicit ennumeration the answer would be deduced as $(10 + 5 + 1)/32 = 0.5$. The first method shows that the answer is 0.5 for *any* odd number of coins.

(4) If 4 boys and 3 girls take random positions in a queue, what is the probability that all the girls are at the head of the line?
Solution: There are 7! ways of placing the 7 children in a line, because there are 7 ways of picking the first child, 6 ways of picking the next, and so on. In a random selection, these 7! alternatives are equally likely. Among these possibilities, there are $3! \times 4!$ ways of arranging matters so that all the girls come before the boys, because there are 3 ways of picking the first girl, 2 ways of picking the second, leaving only one way of picking the third girl; and then there are 4 ways of picking the first boy, etc. So the answer to the question follows from the general idea: probability equals the number of desired outcomes divided by the number of equally likely outcomes, i.e., $(3! \times 4!)/7! = 1/35$.

(5) From balls numbered 1, 2, 3, 4, 5, 6, two are chosen at random.

   (a) How many possibilities are there, if order does not matter?

(b) If a random pair of numbers from 1 to 6 is now guessed, in how many ways can:

    (i) the guessed pair agree with the previously chosen one?

    (ii) one member of the guessed pair agree and one disagree?

    (iii) both guessed numbers disagree with the previously chosen ones?

(c) Check your answers in (b) for consistency with (a).

(d) What is the probability of guessing at least one of the chosen numbers correctly?

Solution:

(a) The number of ways of choosing 2 items in either order from 6 is

$$\binom{6}{2} = \frac{6 \times 5}{2 \times 1} = 15.$$

(b)   (i) Both can be guessed correctly in 1 way.

    (ii) Either of 2 can be guessed correctly, and there are 4 possibilities for the incorrect guess, leading to a total of 8 ways.

    (iii) Both are guessed incorrectly by picking any 2 of the 4 unchosen balls in

$$\binom{4}{2} = \frac{4 \times 3}{2 \times 1} = 6 \text{ ways.}$$

Note: These results can also be obtained by explicitly counting out the fifteen possibilities as 12, 13, 14, 15, 16, 23, 24, 25, 26, 34, 35, 36, 45, 46, 56. Suppose, to be definite, that 12 was the chosen pair. (Any one would do equally well for the argument.) Then, in the list just given, the first corresponds to (i), the next eight to (ii), and the remaining six to (iii), verifying the answers given.

(c) From (b) the total number of mutually exclusive possibilities which exhaust all outcomes is $1 + 8 + 6 = 15$, which is consistent with the answer in (a).

(d) The probability of guessing at least one correctly is the sum of the probabilities of (i) and (ii), or equivalently unity minus the probability of (iii). Either way, the answer is

$$\frac{9}{15} = \frac{3}{5}.$$

(6) An epidemic of a new variety of flu has hit a city. The key symptom of the ailment is stomach pain. At the moment, one person out of ten is known to be suffering from the flu. The probability of someone having stomach pains is 0.7 if he or she has the flu; it is 0.2 for someone who does not have the flu.

(a) What is the probability that a person selected at random has stomach pains and the flu?

(b) What is the probability that a person selected at random has stomach pains but no flu?

(c) A person visits a physician complaining of stomach pains. What is the probability that he or she has the flu?

Solution: There are four mutually exclusive possibilities here: (1) Flu and pains; (2) flu and no pains; (3) no flu and pains; and (4) no flu and no pains. The probabilities of these four events, from the information given are: (1) 0.1×0.7 or 7%; (2) 0.1×0.3 or 3%; (3) 0.9×0.2 or 18%; and, (4) 0.9×0.8 or 72%. [These four probabilities add up to 100% as they must.] The answer to (a) is thus 7%, and to (b) is 18%. In (c) we are given the additional information that either case (1) or case (3) has occured. So we have to consider the relative likelihood of these two outcomes, and the answer to the question is $\frac{7}{7+18} = \frac{7}{25}$, or 28%. [Note: This kind of reasoning, in which probabilities are modified on the basis of known information, is sometimes called Bayesian after the Reverend Thomas Bayes (1702–61). In the present example the information given has clearcut consequences. For a less obviously correct, and thus more controversial, use of this kind of reasoning see solved problem (1) at the end of Chapter 8.]

(7) Four party-goers have stayed till the bitter end. Their four coats hang in the hallway. On leaving they take coats at random. What is the probability that at least one has his or her coat? [Note. This problem is most directly done by counting out the possibilities. Various more clever, but not more correct, methods exist.]

Solution: There are $4! = 24$ ways in which the coats can be shuffled. Let's call the coats A, B, C, and D, and the partygoers *A*, *B*, *C*, and *D*. In the following explicit listing, suppose that *A* takes the first coat, *B* the second, and so on.

ABCD	BCDA	CDAB	DABC
ACDB	BDAC	CABD	DBCA
ADBC	BACD	CBDA	DCAB
ACBD	BDCA	CADB	DBAC
ABDC	BCAD	CDBA	DACB
ADCB	BADC	CBAD	DCBA

By comparing each of the above entries with the ordered listing of partygoers *ABCD* one verifies that there are 15 cases in which at least one of the partygoers has their own coat. The required probability is thus $(15/24) = (5/8)$. [This is a special case of what is called Monmort's problem. The counting of cases systematically turns out to be quite subtle. The answer can be expressed as a series: $1 - (1/2!) + (1/3!) - (1/4!)$, and in this form can be generalized to an arbitrary number of people and coats. Here, then is an example of a problem that is simple to state, and not so easy to solve in general.]

(8) On a TV show, three shoe boxes are put in front of you. You are told that two are empty, and one contains a gold watch. The rules are as follows. You are to pick one box, but not open it. The host, who knows which box contains the watch, will then open one of the two other boxes and show you that it is indeed empty. You will then have the option of changing your choice. Should you? Analyze carefully.

Solution: There are three cases, listed below. [E means empty, G means gold watch.]

	Box 1	Box 2	Box 3
Case I	E	E	G
Case II	E	G	E
Case III	G	E	E

As the listing shows, on your first choice you have a probability 1/3 of getting the watch no matter which box you choose. Let's suppose that you choose Box 1. In Case I, the host *must* now open Box 2. Switching to Box 3, gets you the watch. In Case II, since the host must open Box 3, switching to the remaining Box 2 again makes you a winner. In Case III you were lucky to begin with and you lose by switching. Switching is, however, a good strategy, because by so doing you change your chances of getting the watch from the original 1/3 to 2/3. The reasoning and the conclusion are the same if you had chosen Box 2 or Box 3 at the start.

The last problem is the sort of 'brain-twister' that has the effect of convincing most people that the subject of probability is intrinsically difficult. It is true that it can be difficult to count cases. The main text of this chapter was, by contrast, designed to show that the *basic ideas* – of what is meant by the probability of an elementary event, and of how the probabilities of compound events are then obtained – are quite straightforward.

3

Normality and large numbers

... to be abnormal is to be detested.
Ambrose Bierce

The subject of probability is made particularly interesting and useful by certain universal features that appear when an experiment with random outcomes is tried a large number of times. This topic is developed intuitively here. We shall play with an example used in Chapter 2 and, after extracting the general pattern from the particular case, we shall infer the remarkable fact that only a very small fraction of the possible outcomes associated with many trials has a reasonable likelihood of occurring. This principle is at the root of the statistical regularities on which the banking and insurance industries, heat engines, chemistry and much of physics, and to some extent life itself depend. A relatively simple mathematical phenomenon has such far reaching consequences because, in a manner to be made clear in this chapter, it is the agency through which certainty almost re-emerges from uncertainty,

The binomial distribution

To illustrate these ideas, we will go back to rolling our hypothetical fair dice. Following the example of the dissolute French noblemen of the seventeenth century, one of whose games we analyzed in such detail in the last chapter, we shall classify outcomes for each die into the mutually exclusive categories 'six' and 'not-six', which exhaust all possibilities. If the repeatable experiment consists of rolling a single die, the probabilities for these two outcomes are the numbers $1/6 = 0.16667$ and $5/6 = 0.83333$. These two probabilities are illustrated in the upper left hand corner of Fig. 3.1, where two blocks whose heights are the probabilities for the two outcomes have been drawn side by side. Such a figure is called a histogram, and the collection of two numbers is a (primitive) probability 'distribution', which word in our context denotes the collection of probabilities associated with any set of mutually exclusive outcomes that exhaust all possibilities. Each one of the elements of a probability distribution is a number between zero and one, and the sum of all the elements must add up to unity. (Rule (1), if you wish, of Chapter 2.) Now, let us examine the probability distributions for the repeatable experiment of rolling more than one die, when the outcomes are classified by enumerating 'sixes' and 'not-sixes'. For four dice we did this in great detail in the last chapter, where we calculated as fractions the five probabilities that are listed to five decimal places in Table 3.1. You will verify that this distribution is plotted as a histogram in the lower left hand corner of Fig. 3.1.

I invite you to calculate the distribution for the repeatable experiment of rolling two dice, and to verify that this is correctly plotted in the upper right hand corner of Fig. 3.1.

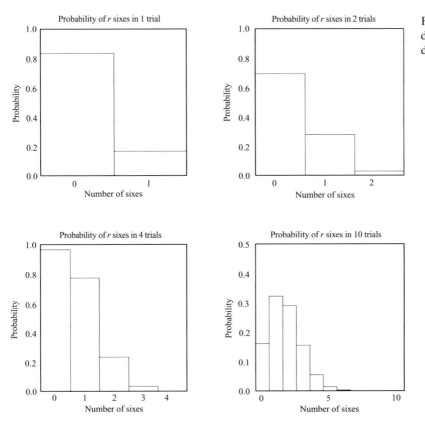

Fig. 3.1. Probability distributions described in the text.

The remaining drawings in Figs. 3.1 and 3.2 show probability distributions in which the repeatable experiment is N rolls of a fair die, and the outcomes are classified according to the total number of sixes that occur. In the different plots, N is 10, 40, 80, 160, and 320 respectively. It is well within our powers now to deduce the general formula used to calculate these distributions. In fact, we can be quite general and assume that the experiment consists of N trials of any two-outcome experiment with probabilities p for 'success' and $(1 - p)$ for 'failure'. (p is $1/6$ in the special case.) What is the probability of exactly r successes in N trials? The method of reasoning is the same as that used for $N = 4$ in Chapter 2. The answer is a product of three factors:

(1) the number of ways in which the r successes can occur during the N trials, which is precisely the number of combinations of N things taken r at a time;

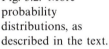

Fig. 3.2. More probability distributions, as described in the text.

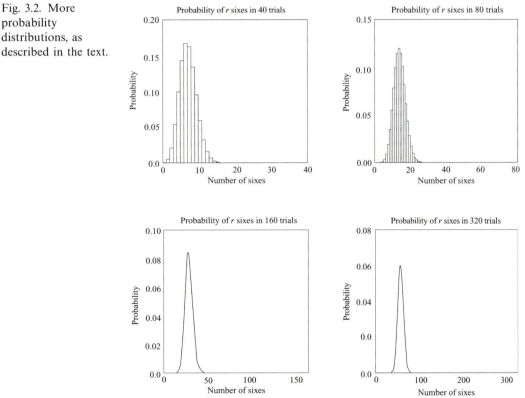

(2) the probability of *r* successes, which we know from previous work
is the probability of one success multiplied by itself *r* times; and

(3) the probability of $N - r$ failures, or, equivalently, the probability
of one failure multiplied by itself $N - r$ times.

Putting these three factors together, and calling the probability in
question $P(r)$, read as 'P of r,' we obtain the result

$$P(r) = \binom{N}{r} p^r (1 - p)^{N-r}. \tag{3.1}$$

The collection of numbers $P(r)$ for $r = 0, 1, 2, \ldots N$ is called the
binomial distribution, the prefix *bi-* referring to the fact that the
elementary experiment that is tried *N* times has *two* possible outcomes.
You may wish to verify that for $N = 4$, $p = 1/6$, and $n = 0, 1, 2, 3, 4$,
one recovers the numbers calculated in the last chapter and listed as
a probability distribution in Table 3.1.†

Equation (3.1) with $p = 1/6$ was used to plot the graphs in Figs. 3.1
and 3.2, and there is no better way to make the formula speak than to
study these curves. Note first of all that the probability is only defined
for *integer* values of the independent variable, so that the correct way
to plot the distribution is as a collection of points or, alternatively, as
a histogram. For the two largest values of *N*, however, the integers so
densely fill the *x*-axis that it is convenient to connect the points into
continuous curves.

Table 3.1. *Illustration of a simple*
probability distribution.

r Number of sixes	$P(r)$ Probability
0	0.48225
1	0.38580
2	0.11574
3	0.01543
4	0.00078

† It is worth emphasizing that we are enumerating outcomes by putting together those corresponding
to a given total number of successes. This grouping is achieved by adding up the number of ways
in which the *r* successes can occur in *N* trials, and is the origin of the combinatorial factor $\binom{N}{r}$ in
(3.1). Our choice of outcomes is thus 'coarse' in the sense of not distinguishing between the different
ways in which *r* successes and $N - r$ failures can be ordered. Later in the book, in connection with
applications to physics, we shall call such a coarse description 'macroscopic,' and the outcomes so
enumerated 'macrostates,' to contrast it with a 'microscopic' description in which the outcomes are
finely distinguished 'microstates.'

Looking at the eight plots in Figs. 3.1 and 3.2, one observes three striking facts:

(a) As N increases, the lopsided distributions characteristic of small N become more and more symmetrical, acquiring a characteristic bell-like shape.

(b) Also as N increases, the distributions begin to peak at the same fractional distance along the x-axis, and a closer examination shows that the peak occurs at $r = (N/6)$.

(c) There is a systematic trend towards narrower peaks as N is increased.

Each of these observations must be, and indeed is, a mathematical consequence of the general form (3.1). One can make them more or less plausible, but there is something here that is not obvious to the unaided intuition. The first remark can be partly understood by noting that the asymmetry of the distribution for a small number of trials is due to the fact that in our example the probability for failure on a single trial, namely $\frac{5}{6}$, is much bigger than the $\frac{1}{6}$ probability for success. By the time one has 10 trials, as in the lower right hand corner of Fig. 3.1, the fact that there are many ways of achieving a given number of successes – by spreading them around among the total number of trials – when this number is not near 0 or 10 begins to become important. The peaked structures in Fig. 3.2 are due to a competition between the coefficient $\binom{N}{r}$, which is largest for r near $\frac{N}{2}$, and the factor $p^r(1-p)^{N-r}$ which is largest for small r when $(1-p)$ is bigger than p. There is also something quantitative contained in remark (a). This I do not know how to explain in words: the explanation will come in further analysis.

The observation (b) about the location of the peak can be understood in simple language. If p is the probability of success on a single trial, it is natural to expect that the average number of successes in N trials will be $p \times N$. Thus, when the distribution is peaked it should peak as observed at $r = (N/6)$.†

The third remark is the effect I have referred to more than once. It is that <u>the number of outcomes for which there is an appreciable probability becomes a smaller and smaller fraction of all possible outcomes as the number of trials increases.</u>

This remark has a quantitative side to it which does not have to be buried in mathematics, so I now want to spend some time playing

† Below, the meaning of the word average is made more precise, and we shall see that this statement about averages is in fact true for every one of the curves in Figs. 3.1 and 3.2.

with the calculated distributions to see if we can make our observation more sharp.

How, then, does the region of appreciable probabilities in the binomial distribution vary with the number of trials? One can attempt to answer this question by looking at the computer-generated tables of numbers from which the curves in Fig. 3.2 were plotted. First, we have to decide on a quantitative definition of the width of a peaked curve like the ones we are considering. There are many possible criteria. We shall pick one that makes particularly good sense for a probability distribution. Let us identify a region of outcomes surrounding the peak which contains a substantial fraction of the probability, and, to be definite, take that fraction to be 99%. We can ask the computer to tell us the range of rs symmetrically surrounding the peak which contains 99% of the probability. For $N = 40$ the answer comes out that this region lies between $r = 1$ and $r = 13$. So, according to our criterion, the width for $N = 40$ is $13 - 1$ or 12. Continuing in the same way we obtain the results given in Table 3.2.

On close observation, Table 3.2 suggests something interesting. We notice that when the number of trials is quadrupled, as in going from 40 to 160, or from 80 to 320, the width doubles. This is exactly what would happen if the width were growing like the *square root* of N. If this observation withstands scrutiny, and we shall see that it does, it begins to answer in a quantitative way the question of how the range of likely outcomes, considered as a fraction of the number of trials, decreases as the number of trials increases.

To explore a little further the proportionality of the width of the probability distribution for N trials of a two-outcome experiment and the square-root of N (which, I remind you, is often written $\sqrt{N}$) let us return to coin tossing, a binomial process with equally likely outcomes, so that p and $1 - p$ in (3.1) are both $\frac{1}{2}$. I programmed a computer

Table 3.2. *99% probable range of outcomes in plots of Fig. 3.2.*

Number of trials	Width of distribution
40	12
80	17
160	24
320	34

to calculate the 99% width, defined exactly as above, for N tosses of a fair coin using (3.1). The results are listed in Table 3.3. The last column shows that the formula $2.6 \times \sqrt{N}$ does quite well.†

Before going on to firm up some of the things we are discovering empirically, it is appropriate to reflect on their implications. For coin-tossing, the probability distribution is symmetrical about $n = \frac{1}{2}N$. Thus, the formula we have just made plausible says in words that it is 99% probable that in N tosses of a fair coin the number of heads will fall in the range $\frac{1}{2}N \pm 1.3 \times \sqrt{N}$, when N is roughly 20 or greater. In particular, for $N = 100$ this formula gives the 99% probable range as 50 ± 13. This is a remarkable statement. Although it does not deal in certainties, it contains a *quantitative* assessment of likelihood, or, to use a term favored by physicists, 'uncertainty.'

What I meant when I talked about certainty almost re-emerging from uncertainty should now be becoming clearer. When an experiment that has no preference between two outcomes is tried 100 times, there is a 99% probability that one of the outcomes will occur between 63 and 37 times. To the question of what that means, you also have the precise answer. Consider the 100 trials as a repeatable experiment, and classify the outcomes of this larger experiment according to whether they do or do not fall into the 99% probable group just described. The new experiment is also a binomial process, but now with 'success' having a probability $p = 0.99$. Thus, when the 100-trial experiment is tried over and over again the number of 'failures' will be very small, one in a hundred trials on average. If there seems to be something elusive in this chain of reasoning, it is the will-o'-the-wisp of absolute certainty, which must necessarily always elude us when we are dealing with random events.

Table 3.3. *Range of 99% probable outcomes for N tosses of a fair coin.*

Number of trials	Width of distribution	$2.6 \times \sqrt{N}$
16	10	10.4
32	15	14.6
64	21	20.8
128	30	29.4
256	42	41.6
512	59	58.8

† This formula does not work for the results in Table 3.2. The question of how the width depends on p will be answered in a later section of this chapter.

The art of the statistician, and I am not one, is to construct reasonably sure tests of hypotheses from a fraction of all the potentially
available data. This involves a number of subtleties which often do not
arise when statistical ideas are applied to the physical world, because
many parts of the world of physics are made up of astronomically
large numbers of identical and almost independent repetitions of a
few basic building blocks.

The mean and the standard deviation

To see deeper into the features of the binomial distribution that
we have discovered empirically, there is no way of avoiding some
mathematical analysis. We shall find that it is not too complicated
to learn about averages and widths by manipulation of the basic
mathematical expression (3.1). This is taken up in the present section.
Substantiating the observation that there is something universal about
bell-shaped distributions is harder, so in the next section I shall
resort to pulling the answer out of a magician's hat, thereby, I hope,
persuading you to swallow the pill containing a little mathematics that
I have prepared as Chapter 5.

A complicated random process, an example of which is many trials
of a simple random experiment, has many possible outcomes. We have
called the corresponding collection of probabilities a 'distribution.' If
a distribution is sufficiently peculiar, so that a graph of it has many
bumps and wiggles, the only way of conveying the information it
contains is to list all its entries in the manner of Table 3.1. Fortunately,
many distributions that occur naturally, for example the binomial
distribution corresponding to a large number of independent trials,
have a simpler structure. These are well described by a few parameters.
For example, in Fig. 3.2 one's eye picks out the position and width
of the single peak as the natural way to describe each curve. A
convenient mathematical way to assign numbers to these descriptive
characteristics, and a calculation of these numbers for the binomial
distribution, is the topic of this section. I ask you to bear with a certain
number of definitions and a certain amount of new terminology here:
they will have their rewards.

A natural measure of where the main weight of a distribution is
located can be found in a generalization of the everyday notion of
'average.' To keep things simple, imagine a room full of people all
of whom are 5 ft 11 in, 6 ft, or 6 ft 1 in tall. Their average height,
you would agree, is obtained by adding together their heights and

dividing by their number. An entirely equivalent way of calculating
the same average would be: first, to calculate the fractional number in
each category; then, to multiply each of the fractions by the common
height of the members of that category; and, finally, to add the three
products. If this is not clear, please try it both ways imagining that
there are 5, 30, and 5 members of the respective categories. Either
way you do it, the average will come out to be 6 feet. For example,
in the first way of calculating there will be 30 entries of 6ft added
together, and the final sum would be divided by 40. But this part of
the calculation gives $\frac{30}{40} \times 6$ ft, which is the contribution of people 6 ft
in height to the second way of calculating the average.

The 'mean' of a distribution $P(r)$, to which we shall attach the
Greek symbol μ, pronounced 'mu,' is defined by the equation

$$\mu = 0 \times P(0) + 1 \times P(1) + 2 \times P(2) \ldots N \times P(N)$$
$$= 0 + 1P(1) + 2P(2) + \ldots NP(N). \tag{3.2}$$

[The second line above is the same as the first; the multiplication signs
have been omitted using the convention in algebra, already used in
(3.1) that juxtaposition means multiplication.] In (3.2) the dots mean
that one is to add all terms of the form $rP(r)$ for r intermediate between
2 and N, and we have assumed that r, like the number of successes
in the N trial binomial distribution, can take on the numerical values
$r = 0, 1, \ldots N$. The mean describes the distribution in exactly the same
way that the average height described our room full of people.

The average height of a collection of individuals tells us something
significant, but by no means everything, about the group. A collection
consisting of an equal number of jockeys 5 ft tall and basketball-
players 7 ft tall has the same average height as our previous one,
which silly example illustrates that it would be informative to know
how much *variation* there is around the average height. A first try at
a numerical way of conveying this information might be to subtract
the average height of the group from each person's height, and to
average all the resulting numbers. Unfortunately, because there are as
many positive as negative deviations, the end result of this procedure
is the uninformative number zero. Since the squares of numbers
both positive and negative are always positive, a way out of this
embarrassment is to average the squares of the differences of each
person's height from the average height. This quantity is called the
mean square deviation or, equivalently, the 'variance,' and we can
illustrate its calculation by working it out for our two hypothetical
collections of people with different heights. In the first collection, the
middle group – of people 6 feet tall – has exactly the average height;

it therefore plays no role in the calculation of the variance. Working out the contribution of the other two groups, each of which in our example contains (1/8)th of the total number, we get $(-1 \text{ in})^2 \times (1/8) + (+1 \text{ in})^2 \times (1/8)$ which is equal to 1/4 of a square inch. Now one usually associates 'square inches' with an area, and we are not calculating the area of anything here. A number which measures deviations from the average and comes out in inches, i.e. has the same 'dimensions' as the objects being considered (heights in our case), is the square root of the variance. This is called the <u>root mean square deviation</u> or the <u>standard deviation</u>, which for the group we are considering is $\sqrt{1/4} = \frac{1}{2}$ inch. For our second hypothetical collection, of jockeys and basketball players, one verifies that the standard deviation is 1 foot. So you see that our two groups having the same average height are well distinguished by their quite different standard deviations.

Any probability distribution can, in the same way, be assigned a standard deviation. We shall attach the Greek symbol σ, pronounced 'sigma' (or 'lower case sigma'), to this quantity. The formal definition of the variance or square of the standard deviation is then:

$$\sigma^2 = (0 - \mu)^2 P(0) + (1 - \mu)^2 P(1) + \ldots + (N - \mu)^2 P(N). \quad (3.3)$$

The dots again indicate terms of the form $(r-\mu)^2 P(r)$ for r intermediate between 1 and N, and symbols placed side by side are, as in the second form of (3.2), to be multiplied together.†

Since one often encounters sums like those in (3.2) and (3.3) it is convenient to have a more compact notation for them. One writes

$$\mu = \sum_{r=0}^{N} r P(r) \qquad\qquad (3.4)$$

which is read as 'mu equals the <u>sum</u> from r equals zero to r equals N of r multiplied by P of r.' By the same token, we write

$$\sigma^2 = \sum_{r=0}^{N} (r - \mu)^2 P(r). \qquad\qquad (3.5)$$

These two equations are nothing more or less than a shorthand for the previous two. In words, one is to take whatever sits on the right of the Greek letter Σ ('capital sigma'), to calculate it successively for each integer r starting at the value written below the symbol and ending at the value written above it, and to add together the results of the

---

† Another entirely equivalent but sometimes more convenient expression for the standard deviation will be found in an appendix to this chapter.

successive calculations. As examples consider

$$\sum_{r=0}^{4} r = 0 + 1 + 2 + 3 + 4 = 10, \tag{3.6}$$

and

$$\sum_{r=0}^{4} r^2 = 0 + 1 + 4 + 9 + 16 = 30. \tag{3.7}$$

How are the features that our eyes picked out in Figs. 3.1 and 3.2 related to the means and standard deviations of the distributions plotted there? To answer this question, we have to work out these quantities for the binomial distribution (3.1). At first sight, this seems like a formidable task, but, since we have already learned the power of working out special cases, let's approach it step by step. Try $N = 1$, i.e. the experiment of one trial of a two-outcome process. Now we have the distribution consisting of the two numbers $P(0) = 1 - p$ and $P(1) = p$, which was plotted for $p = 1/6$ in the upper left hand corner of Fig. 3.1. You will verify that the mean of this distribution, call it μ_1, is p, and the standard deviation, call it σ_1, is $\sqrt{p(1-p)}$.† [Don't be lazy. Try it.]

Now, I am going to show you that the $(N + 1)$-trial mean, μ_{N+1}, is related to the N-trial mean, μ_N, through the formula

$$\mu_{N+1} = \mu_N + \mu_1. \tag{3.8}$$

From this it follows that $\mu_2 = 2\mu_1$. Proceeding in this way, integer step by integer step, we see that

$$\mu_N = N\mu_1. \tag{3.9}$$

Here is a way of proving (3.8). I'm going to do it for $N = 2$, so that $N + 1 = 3$, but in such a way that the generality of the method should be evident. In Table 3.4 are listed the outcomes and probabilities for three binomial trials. In Column 1, the outcomes have been listed according to whether there were successes (s), or failures (f), in the first, second, and third trials. Counted this way, there are eight possible outcomes. The probabilities of the different outcomes are listed in Column 2, and the number of successes in Column 3. For reasons that will become clear immediately, the successes have been recorded as a sum. The first entry, r_2, shows the number of successes in the first two trials; the second, r_1, shows whether the last trial resulted in a

---

† Note by extracting the common factor $p(1-p)$ that $(-p)^2(1-p)+(1-p)^2p = p(1-p)[p+(1-p)] = p(1-p)$

success or a failure. Unnecessarily complicated though the table may
seem, you will have to agree that it is a perfectly correct listing of the
probabilities for all possible results of three trials of a two-outcome
experiment.

Several interesting properties of the distribution can be inferred
easily from this way of looking at it. First, let us see that the
probabilities add up to one, thereby verifying that we are dealing
with a legitimate distribution. Notice that the listing in Column 1
is such that all entries but the last are identical for each pair of
lines starting from the top. The last entry alternates between success, which leads to a multiplicative factor of p in Column 2, and
failure, with a corresponding multiplicative factor of $1 - p$. Since
$p + (1 - p) = 1$, a pairwise addition – i.e. the first and the second, the third and the fourth, and so on – of the 8 entries in the
second column of the table gives the 4 probabilities for the outcomes
of two trials of a binomial experiment, when the order of successes
and failures is, as here, distinguished. We see, therefore, that the
sum of the probabilities for three trials is equal to the sum of the
probabilities for two trials, which, continuing the process, is equal
to the sum of the probabilities for one trial, or unity. A little reflection will convince you that this argument is general: if we had
started in our mind's eye with the 2^N strings of ss and fs for N trials, and arranged these strings analogously to those in Table 3.4, a
sequential pairwise addition would show that the sum of the proba-

Table 3.4. *Outcomes and probabilities for three trials of a*
binomial experiment.

Column 1	Column 2	Column 3
Outcomes	Overall Probability	Number of Successes
		$r_2 + r_1$
s s s	ppp	2 + 1
s s f	$pp(1-p)$	2 + 0
s f s	$p(1-p)p$	1 + 1
s f f	$p(1-p)(1-p)$	1 + 0
f s s	$(1-p)pp$	1 + 1
f s f	$(1-p)p(1-p)$	1 + 0
f f s	$(1-p)(1-p)p$	0 + 1
f f f	$(1-p)(1-p)(1-p)$	0 + 0

bilities is one. The conventional way to describe this crucial property of a legal probability distribution is to say that the distribution is 'normalized.'

Let's use our table to see how the mean of the binomial distribution varies with the number of trials. To do this for three trials, we have – following the general rule (3.2) or (3.4) for calculating the mean – to multiply together the entries in each row of the last two columns and then to add up all these products. We shall do this in three steps. First, take the number before the plus sign in the 'Number of Successes' column, multiply it by the probability on its left, do the same for each row of the table, and add all the results; then, perform the analogous operations with the numbers after the plus sign; and, finally, add the two sums. No error is made in this process because, to take the first line as an example, $(2 + 1)p^3 = 2p^3 + 1p^3$. More important, something has been achieved by this tactic. The first number in the last column is identical in each pair of lines starting from the top, for the very good reason that it is the number of successes in the first two trials, and this number is the same in each pair of lines. Thus, adding the lines pairwise exactly as in the previous paragraph, and using the property exploited there, we find that the first of our two sums is the mean number of successes in *two* trials. The second number in column 3, because of the way in which we have organized things, alternates between 1 and 0. The second sum in our three-step procedure is then seen to be $p\times$ (the sum of all the probabilities for two trials, a sum which we know to be 1). In short, we have very explicitly proved (3.8) for $N = 2$, and done it in such a way that the generalization to arbitrary N requires only the kind of thinking contained in the last paragraph. In fact, we wrote μ_1 in this equation instead of p, the value of the single trial mean in the example we have just done, because the industrious reader will be able to generalize the argument to the case of N independent trials of any random process, with μ_1 being the mean of the distribution for a single trial.

If the argument just given seems involved, that is only because it takes many words to explain what the mind perceives more quickly when it sees through the verbiage. If you have succeeded in doing this, you have understood, perhaps in more detail than you ever wanted, why the peaks in the curves in Fig. 3.2 behave the way they do. Because the eye does not instinctively pick out the mean from a drawing of an asymmetrical distribution, we did not notice in the histograms in Fig. 3.1 that the means in each case are $N/6$. However, for a symmetrical distribution with a single peak, the mean and the

peak coincide. So we understand that the peaks 1/6th of the way out
are a reflection of our theorem (3.9) for a case in which $\mu_1 = \frac{1}{6}$.

There is a similar theorem for standard deviations. The connection
between the standard deviations for trial sizes of $N + 1$, N, and 1 is

$$\sigma^2_{N+1} = \sigma^2_N + \sigma^2_1 \tag{3.10}$$

from which it follows, as (3.9) followed from (3.8), that $\sigma^2_N = N\sigma^2_1$, or,
taking the square root of both sides, that

$$\sigma_N = \sqrt{N}\sigma_1. \tag{3.11}$$

This is very interesting: (3.11) is nothing else than our empirically
discovered result that the width grows like the square root of the
number of trials!

Equation (3.10) can also be proved for $N = 2$ using our table. We
have to calculate (3.5), by which we mean the sum (3.3), for the entries
in Table 3.4. Following the prescription, we have to subtract the
mean of the distribution from each entry in the last column, square
the resulting differences, and average the squares using the entries in
the second column. [Note that I am using the word 'average' in a
manner consistent with earlier usage. The mean was introduced as
a word to describe the average of r, as spelled out in (3.2). By the
same token, the variance defined in (3.3) is the 'average' of $(r - \mu)^2$.]
Now, we know the mean for the three-trial distribution from the work
we have just done: $3p$. It is, however, convenient and in the spirit
of the way in which we have been reasoning here to write the mean
for three trials as the sum of the mean for two trials, $\mu_2 = 2p$, and
the mean for one trial, $\mu_1 = p$. In the table, we have labeled the
first entry in Column 3 r_2 and the second entry r_1, these two symbols
standing for the listed numerical values in the eight rows of the table.
So, the quantity we have to average is, after a trivial rearrangement,
$(r_2 - \mu_2 + r_1 - \mu_1)^2$. It is now convenient to use a result from elementary
algebra: $(a + b)^2 = a^2 + 2ab + b^2$. Taking a to be $(r_2 - \mu_2)$ and b to be
$(r_1 - \mu_1)$, we see that

$$(r_2 - \mu_2 + r_1 - \mu_1)^2 = (r_2 - \mu_2)^2 + (r_1 - \mu_1)^2 + 2(r_2 - \mu_2)(r_1 - \mu_1). \tag{3.12}$$

The average we are calculating is thus the sum of the averages of the
three terms displayed on the right side of this equation. The first of
these, $(r_2 - \mu_2)^2$, does not involve the entries in Column 3 after the
plus sign. Making a pairwise addition of the rows, we see that the
average of this term is the variance for *two* trials. For the average
of $(r_1 - \mu_1)^2$, add together all the rows for which $r_1 = 1$ and then
all the rows for which $r_1 = 0$, to see that this is the variance for one

trial. Finally, consider the cross term $(r_2 - \mu_2)(r_1 - \mu_1)$. Its average is zero, because it involves the averages of deviations from the mean. Explicitly, the average of $(r_1 - \mu_1)$ over any pair of lines in which r_2 is constant involves $(0 - p)(1 - p) + (1 - p)p$ which is a way of writing the number zero. So we are done; we have proved (3.10) for $N = 2$ in a way that can be generalized to arbitrary N, and, therefore, (3.11).

This has been a heavy dose of algebra, but something has been achieved. We have associated two descriptive numbers with a probability distribution: the mean, μ, which tells where the weight of the distribution is centered; and the standard deviation, σ, which is a quantitative measure of deviations from the mean. We found that for N independent trials of a random experiment the mean is $N \times$ (the single trial mean), and the standard deviation is $\sqrt{N} \times$ (the standard deviation for a single trial). (I am stating generally a result that we proved in all detail in a special case, because the proofs contain the seeds of a general argument.)

Our analysis has explained some, but by no means all, of the observations we made by looking at drawings of the binomial distribution, and it has raised at least one new question. Perhaps the most intriguing observation, that a plain and featureless distribution could give birth to a series of elegant bell-shaped curves, has yet to be understood. And although we have shown that a natural measure of the width of the distribution (the standard deviation) scales with the square root of the number of trials, we are left with the question of how the standard deviation is related to the width defined as the range of outcomes containing 99% of the probability. Evidently, the two are not identical. For N tosses of a fair coin we found the 99% range to be given approximately by $1.3\sqrt{N}$ on either side of the mean, whereas the standard deviation for this case, $p = \frac{1}{2}$, is $\frac{1}{2}\sqrt{N}$. If our observation has some generality to it, it would seem to be saying that 99% of the probability is contained in about 2.6 standard deviations on either side of the mean. To test if this is roughly true, we can try it out on Table 3.2 in which we listed the 99% probable range for the number of sixes on N rolls of a fair die. The hypothesis would predict the range to be $2 \times 2.6 \times \sqrt{N \times (1/6) \times (5/6)}$. [Do you agree?] This formula gives the numbers 12.3, 17.3, 24.5, and 34.6, to compare with those in the last column of the table, i.e 12, 17, 24, 34. Not bad! In fact, there *is* something general buried here, but to dig it out we are going to need more mathematical tools. So I shall tell you what we are going to find when we are better equipped two chapters ahead.

The normal distribution

The 'normal probability distribution,' also called the 'Gaussian distribution,' is a curve depending on a continuous variable, call it x, and two numbers which are usually named μ and σ. The curve is bell-shaped, the area under it is unity, the peak is at $x = \mu$, and the standard deviation – as generalized in a natural way to apply to a continuous curve – is σ. There is a simple mathematical formula for the curve, but it is premature to write it out now. Instead, in Fig. 3.3, this 'function of x', to use the mathematical term, is plotted for $\mu = 40$ and $\sigma^2 = 20$. The histogram on the same plot gives the probabilities for various numbers of heads on 80 tosses of a fair coin, i.e., a binomial distribution with, according to what we have just learned, a mean of $80 \times (1/2) = 40$ and a variance of $80 \times (1/2) \times [1 - (1/2)] = 20$. You will notice that the curve at integer values of x is very closely equal to the height of the corresponding column of the histogram. Although there are tiny differences, too small for the eye to detect, in the far wings, the message of the figure is that the binomial distribution for a large number of trials is extremely well approximated by a normal curve with the same mean and standard deviation. This is the root cause of the similar shape of the last three distributions plotted in Fig. 3.2: they are each, to all intents and purposes, normal distributions differing only in their means and standard deviations.

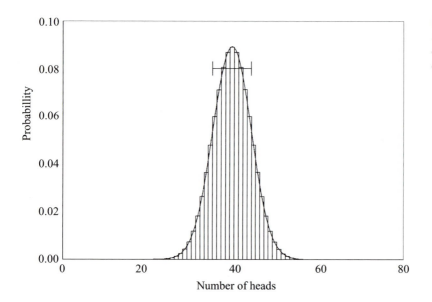

Fig. 3.3. Binomial and normal distributions, as described in the text.

The normal distribution is so well studied and occurs so frequently that its properties are to be found in even rather small volumes of collected mathematical tables. One finds, for example, that a region of 2.6 standard deviations on either side of the mean contains 99.07% of the area under the normal curve. Since the area is the natural generalization of the sum of columns in a histogram, the reason for the 'universality' – to use a word quite fashionable among physicists today – of the connection between the standard deviation and the 99% width has been uncovered.

The point is that after a moderate number of trials of a two-outcome random experiment, or indeed any reasonable random experiment, an approximation to the normal distribution emerges with the same mean and standard deviation as the N-trial distribution. The rule of thumb '2.6 standard deviations on either side of the mean' thus applies generally as an estimate of the 99% probable range after N trials, when N is large enough for the distribution to be bell-shaped. A table of the normal distribution also reveals that 2 standard deviations on either side of the mean contains 95.45% of the area. Thus, 2 standard deviations is a good estimate for the 95% probable width. Similarly, one sees from Table 3.5 that 1 standard deviation on either side of the mean contains approximately 68% of the probability. [This width, 1 standard deviation on either side of the mean, is indicated by the horizontal line in the upper region of Fig. 3.3.]

It is important to re-emphasize that the normal distribution only applies after a reasonably large number of independent trials. For a small number of trials, the question of how likely or unlikely is a given collection of outcomes can only be answered by examining the distribution particular to the case in question.

The laws of averages and large numbers: the central limit theorem

The knowledge we have acquired is interesting, by no means obvious, and useful, an assertion that is best substantiated by examples. First, however, I think I should tell you that you have just learned everything you always wanted to know about what are called the law of large numbers and the central limit theorem. Among other things, these learned codifications, to which we shall come in a moment, contain what is popularly known as the law of averages. The law of averages is the statement that the most likely fraction of successes in a sufficiently large number of trials of a random experiment is the probability of success on a single trial. It may also be appropriate to state what the law of averages is not, though no one who has read this far could share what some writers claim are common misconceptions on this subject. The law of averages does not introduce connections between unconnected events. Misconception about conceptions... If ten baby boys are born one after the other in a certain hospital, that would have to be considered uncommon, having a probability of $(\frac{1}{2})^{10}$, or just under 0.001, on the assumption that boys and girls are equally likely. But an increase in the likelihood of a girl being born next, which is often said to be a common view, does not, of course, come about as a consequence. The law of averages is concerned with precisely the question we have been exploring in this chapter, and would say in the matter of ten births chosen at random that the most likely situation would be five boys and five girls. We are in a position to be a lot more precise than that. We can specify the range in the number of girls to be expected on the basis of some given criterion of likelihood. The numbers we are talking about are too small for the rules of thumb discussed in the last section to work very well, but it is easy enough to calculate that the probability of one girl, or of nine girls, is just under 1%. Adding in the very small chance of zero or ten girls, we conclude that it is 98% probable that somewhere between two and eight girls will be born in a random selection of ten births.†

This example illustrates that for smallish numbers of trials the law of averages by itself is not a very precise predictor. When larger numbers are involved, the fact that the square root of such a number is very much smaller than the number itself begins to work its powerful way, the probability distribution squeezes down around the mean, and the observed fraction of outcomes of a particular sort becomes a better and better estimate of the probability of those outcomes in a single trial. This is the law of large numbers.

The central limit theorem goes further and says that as the number of independent trials becomes larger and larger the distribution of outcomes is better and better approximated, for small percentage deviations from the mean, by the normal distribution.‡ Very little needs to be said on these matters, apart from giving you the names

---

† See solved problem (4) at the end of this chapter for the calculation.

‡ Just how many trials are needed depends on the circumstances. For example with an experiment having a success probability of 0.01, 100 trials is not large enough, because the mean number of successes, 1 in this case, is so small that the correct (binomial) distribution is not yet bell-shaped. See the last solved problem at the end of this chapter.

bestowed on them by higher learning, because at a conceptual level you already have a complete understanding of the central ideas.

Summary

Before going on to fix these ideas in place and demonstrate their power in special cases, let us take a brief look backward at what has already been achieved. We have learned how to assign probabilities to the result of many independent trials of a random experiment, and we have reached some significant conclusions about the region where appreciable probabilities are found. Rather than repeat these conclusions in a general form, let me say that you have understood them if you have fully understood the sentence on page 29: When an experiment that has no preference between two outcomes is tried 100 times there is a 99% probability that one of the outcomes will occur between 63 and 37 times. This is a 26% range. What happens if the number of trials increases? For 10000 trials, the same general formula – 2.6 standard deviations on either side of the mean – gives the 99% range as 5130 to 4870. (Do you agree?) This is a 2.6% range. For 1 000 000 trials, this decreases to 0.26%. As the number of trials increases further, the range of typical 'statistical fluctuations,' to use the words that are used to describe this phenomenon, becomes a yet smaller percentage of that number.

You will agree that these facts are not obvious to the unaided intuition. And yet, they are neither counter-intuitive nor especially hard to grasp. What is surprising is that they are not more widely appreciated.

To close this chapter, I list in Table 3.5, for future reference, a few entries from a table of the normal distribution. The first entry tells how many standard deviations on either side of the mean one is contemplating, and the second tells the probability associated with this range of outcomes.

All of the probability of the normal distribution is evidently concentrated within a few standard deviations of the mean.

Appendix: Another expression for the variance

There is another often more convenient formula for the variance than the one given in (3.5). Using the relation $(r - \mu)^2 = r^2 - 2r\mu + \mu^2$ in this equation, it follows that

$$\sigma^2 = \sum_{r=0}^{N}(r^2 - 2\mu r + \mu^2)P(r)$$

$$= S_1 - 2\mu S_2 + \mu^2 S_3,$$

(3.13)

where

$$S_1 = \sum_{r=0}^{N} r^2 P(r),$$

(3.14)

$$S_2 = \sum_{r=0}^{N} r P(r) = \mu, \quad \text{and}$$

(3.15)

$$S_3 = \sum_{r=0}^{N} P(r) = 1.$$

(3.16)

To understand equations like the ones above, it is necessary to think concretely about what they mean, using, when needed, the explanation given after (3.5) to read them in the 'longhand' in which (3.3) is written. Viewed in this way, it is clear that in going from the first to the second line of (3.13) we have taken out -2μ as a common factor in the second term and μ^2 as a common factor in the third. In (3.15) we note that the sum we have called S_2 is nothing else but the average of r, or the mean μ, of the distribution $P(r)$. In (3.16) we note that the sum we have called S_3 is the sum of all the elements of the distribution $P(r)$ which is unity because it is 'normalized.' These observations show that the second term on the second line of (3.13) is the negative of twice the last one.

Table 3.5. *Normal probability integral. The second column gives the area under the normal distribution curve for the number of standard deviations on either side of the mean shown in the first column.*

Standard deviations	Probability
0.1	0.0797
0.5	0.3829
1.0	0.6827
1.5	0.8664
2.0	0.9545
2.5	0.9876
3.0	0.9973
3.5	0.9995
4.0	0.9999

Using these facts to combine the last two terms in (3.13) leads to the expression:

$$\sigma^2 = \sum_{r=0}^{N} r^2 P(r) - \left[\sum_{r=0}^{N} r P(r) \right]^2. \tag{3.17}$$

A way of saying this final result in words is: the variance is the average of the square minus the square of the average.

In order to check the correctness and utility of (3.17), consider a single trial of a binomial process. Now, the average of the square is $0^2 \times (1-p) + 1^2 \times p = p$, whereas the square of the average is $[0 \times (1-p) + 1 \times p]^2 = p^2$. Thus (3.17) in this case leads to $\sigma_1^2 = p - p^2 = p(1-p)$, a result you had obtained directly on p. 33 by doing a little more algebra.

Solved problems

(1) The infinite array of numbers of which the first few rows are displayed below is called Pascal's triangle:

```
                        1
                    1       1
                1       2       1
            1       3       3       1
        1       4       6       4       1
    1       5      10      10       5       1
1       6      15      20      15       6       1
1       .       .       .       .       .       .       1
```

Note the following facts:

(a) The $N + 1$st row consists of the combinatorial coefficients $\binom{N}{r}$ for $r = 0, 1, \ldots, N$.

(b) Every interior entry is obtained by adding the two entries immediately above it.

Prove, starting from the general form for the combinatorial coefficient, that

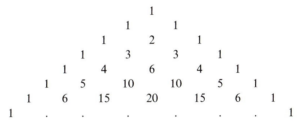

$$\binom{N}{r} = \binom{N-1}{r-1} + \binom{N-1}{r}$$

for $N \geq 2$ and $1 \leq r \leq N - 1$, and that this explains (a) and (b).

[Hint: $(N/r) = 1 + (N-r)/r$]

Solution: Notice that $N! = N \times (N-1)!$, because both sides of this equation consist of the integers from 1 to N multiplied together. Similarly, $r! = r \times (r-1)!$. Exposing the factor N in the numerator, and r in the denominator, allows one to use the hint, as follows:

$$\binom{N}{r} = \frac{N!}{(N-r)! \times r!} = \frac{N}{r} \times \frac{(N-1)!}{(N-r)! \times (r-1)!}$$

$$= [1 + \frac{N-r}{r}] \times \frac{(N-1)!}{(N-r)! \times (r-1)!}$$

$$= \frac{(N-1)!}{(N-r)! \times (r-1)!} + \frac{(N-1)!}{(N-r-1)! \times r!} \quad ,$$

$$= \binom{N-1}{r-1} + \binom{N-1}{r}$$

which proves the proposed identity.

Now the numbers in Pascal's triangle are, evidently, – and this is the observation (a) – the numerical values of the symbols shown below:

$$1$$

$$\binom{1}{0} \qquad \binom{1}{1}$$

$$\binom{2}{0} \qquad \binom{2}{1} \qquad \binom{2}{2}$$

$$\binom{3}{0} \qquad \binom{3}{1} \qquad \binom{3}{2} \qquad \binom{3}{3}$$

$$\binom{4}{0} \qquad \binom{4}{1} \qquad \binom{4}{2} \qquad \binom{4}{3} \qquad \binom{4}{4}$$

$$\cdot \qquad \cdot \qquad \cdot \qquad \cdot \qquad \cdot \qquad \cdot$$

You will verify that the theorem we just proved is an algebraic summary of the statement that in this triangle every interior entry is the sum of the two entries above it, thus explaining the observation (b).

There is another, more intuitive, way to see the connection between Pascal's triangle and the combinatorial coefficients. We have seen that when a coin is tossed N times one can classify the outcomes into 2^N or $N + 1$ outcomes depending on whether one does distinguish the order in which heads and tails come up, as in Table 3.4, or does not, as, e.g., in the histogram in Fig. 3.3. Grouping together the outcomes with the same number of heads, without regard to order, led to the appearance of the combinatorial coefficients. That Pascal's triangle as constructed by the addition rule does precisely this grouping can be seen in the following way. Imagine connecting every point in the array to its two nearest neighbors below it. I assert that the entry at any point is the number of distinct paths by which that point can be reached from the 1 at the apex. For example, the 2 in the third line corresponds to the zig-zag paths right–left and left–right; the 3 below and to the right of this 2 can be reached by the paths right–right–left, right–left–right, and left–right–right. Replace right by 'heads' and left by 'tails' to see that the number of paths is the number of outcomes corresponding to a given number of heads, which is precisely the definition of the combinatorial coefficient. Now the addition rule clearly applies to the number of paths, because paths to a given point come from either its upper left or its upper right, with the consequence that the number of distinct paths to a point is the sum of the number of distinct paths to those two points.

Constructing Pascal's triangle according to the addition rule for the interior entries is thus a, rather laborious, way of calculating the combinatorial coefficients. More important, it provides a way of visualizing the fact that the binomial distribution for coin tossing peaks near its middle, because the numbers in the Nth row when divided by 2^N make up the probability distribution for $N - 1$ tosses of a fair coin. Everything we have learnt in this chapter about the concentration of probability into a region near the mean, is related to the large number of ways in which outcomes in that region can be grouped together when their order is not distinguished.

Of course the force of our observations was due to their quantitative nature, and Pascal's triangle by itself is not sufficient for that purpose.

(2) Calculate (as fractions) the probabilities of obtaining r heads when a fair coin is tossed N times for

$$N = 3 \qquad r = 0, 1, 2, 3$$
$$N = 6 \qquad r = 0, 1, 2, 3, 4, 5, 6.$$

Check that both sets of results are correctly plotted in Fig. 3.4, where the inner scale is for $N = 3$ and the outer one for $N = 6$.

Solution: One can calculate the probabilities as discussed in the chapter, or read them off from Pascal's triangle. Dividing the 4th row by $2^3 = 8$ we have the distribution:

$$\frac{1}{8}, \frac{3}{8}, \frac{3}{8}, \frac{1}{8}.$$

Similarly, dividing the 7th row by $2^6 = 64$, we have the distribution:

$$\frac{1}{64}, \frac{3}{32}, \frac{15}{64}, \frac{5}{16}, \frac{15}{64}, \frac{3}{32}, \frac{1}{64}.$$

These have been plotted in Fig. 3.4. The fact that the region of appreciable probability gets more concentrated near the mean is beginning to emerge even with these very small numbers.

Fig. 3.4. Solution to problem (2)

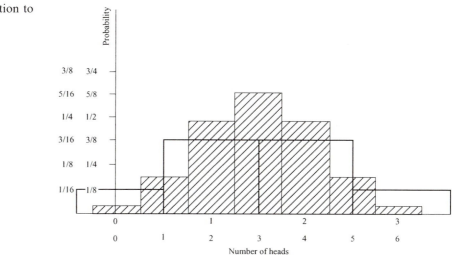

Number of heads

(3) Calculate the probability of obtaining more heads than tails in the throw of 4 unbiased coins. Repeat for 6 unbiased coins. Why is this problem more difficult if the coins are biased?

Solution: The relevant probabilties can again be read off from Pascal's triangle, from the 5th and 7th rows in point of fact. For fair coins, only the middle entry is needed, because in this case more heads than tails is just as likely as more tails than heads. For 4 coins the answer is:

$$\frac{1}{2}\left(1 - \frac{6}{16}\right) = \frac{5}{16}.$$

For 6 coins the answer is:

$$\frac{1}{2}\left(1 - \frac{20}{64}\right) = \frac{11}{32}.$$

The idea is to halve the probability left over when the outcome of exactly as many heads as tails is excluded. The absence of 'symmetry' between heads and tails for a biased coin, would then require an explicit calculation of all outcomes, making it more difficult.

Note that the probability of an exactly even split *decreases* as the number of trials increases. $\left[\frac{6}{16} \text{ is bigger than } \frac{20}{64} = \frac{5}{16}.\right]$ This is not in contradiction to anything we have learnt, but might surprise someone with a naive view of the Law of Averages.

(4) Prove the assertion made on page 40 that, if ten babies are born one after another in a certain hospital, there is 98% probability of there being between 2 and 8 girls amongst them, making the *assumption* that girls and boys are equally likely to be born.

Solution: The probabilities of 0 or 10 girls, under the given assumptions, are each $1/2^{10} = 1/1024$; the probabilities of 1 or 9 are each 10 times as great, because of the different ways in which these outcomes can occur, i.e., because of the combinatorial factor. The total probability of these 4 outcomes is thus $22/1024 \approx 2.2\%$, leaving approximately 98% as asserted. [The symbol $\approx$ means 'approximately equal to.']

Table 3.6. *The probabilities of successes in problem (5).*

Number of successes r	Probability $P(r)$
0	0.3660
1	0.3697
2	0.1849
3	0.0610
4	0.0149
5	0.0029

(5) Consider the experiment consisting of 100 trials of a two-outcome process with success probability 0.01. Can the associated binomial distribution be described by a normal distribution?

Solution: In brief, the answer is *no*. Here, 100 trials are not enough for the universal features of the normal distribution to emerge. With $N = 100$ and $p = 0.01$, the N-trial mean μ_N is $Np = 1$, and the N-trial standard deviation σ_N is $\sqrt{Np(1-p)} \approx 1$. The formulas used, being quite generally true for the binomial distribution, are true here. However, the distribution cannot be approximated by a single symmetrical peaked curve. Applying the formula (3.1) we get, using a calculator, Table 3.6, showing the probabilities for various numbers of successes. The probabilities of no successes and one success are roughly equal, and the probabilities then decline rapidly. [The distribution is rather well approximated by one discussed in the Appendix to Chapter 5.] Roughly ten times as many trials would be needed for a peak to be seen.

4

Examples

. . .lies, damn lies, and statistics.
Mark Twain

Here are a few randomly chosen and occasionally whimsical uses for your new knowledge about the workings of chance. The situations I shall describe all have to do with everyday life. In such applications, the difficulty is not only in the mathematical scheme but also in the frequently unstated assumptions that lie beneath it. It helps to be able to identify the repeatable random experiment and, when many trials are being treated as independent, to be able to argue that they are in fact unconnected.

The examples have been chosen to illustrate the role of statistical fluctuations, because this is the most interesting aspect of randomness for the physical applications to follow. A statistician or a mathematician would choose other examples, but, then, such a person would write a different book.

Polling

Opinion polls are second only to weather forecasts in bringing probability, often controversially, into our daily lives. 'Polls Wrong,' a headline might say after an election. Opinions change, and often suddenly. A pollster needs experience and common sense; his or her statistical knowledge need not be profound. But, there is a statistical basis to polling. Consider the question: 'If 508 of 1000 randomly selected individuals prefer large cars to small, what information is gleaned about the car preferences of the population at large?' If we ignore the subtleties just alluded to, the question is equivalent to the following one. Imagine a large urn (it's always an urn, but you can make it a jar if you wish) containing many, many marbles, all of which are either green or brown. If 508 brown marbles occur in a random selection of 1000 marbles drawn from the jar, what can be said about the ratio of brown to green marbles in the jar as a whole? (Incidentally, since drawing out a thousand marbles would appreciably deplete even an urn holding a few thousand marbles, one could imagine returning each withdrawn marble before drawing the next one, taking appropriate care to stir the container so as not to bias the draw.) A little thought along the lines sketched in the last chapter shows that the result is quite compatible with the *hypothesis* that there are as many green as brown marbles. We can put to use the formulas derived in the last chapter for the mean and standard deviation of a binomial distribution. In the notation used there, the N-trial mean and variance are given by

$$\mu_N = Np, \quad \text{and} \quad \sigma_N^2 = Np(1-p). \tag{4.1}$$

In the present case $N = 1000$ and the hypothesis to be tested is $p = \frac{1}{2}$. For these numerical values, the mean comes out to be 500 and the standard deviation $\frac{1}{2}\sqrt{1000}$, a number somewhat less than 16. A deviation larger than or equal to the observed one of 8 corresponds to 8/16, or approximately 0.5, standard deviations on the high side of the mean. Now, in the discussion at the end of the last chapter we argued that for the moderately large numbers that we are dealing with here the binomial distribution can be well approximated by the normal distribution. From Table 3.5 one reads that the region contained within 0.5 standard deviations on either side of the mean has a probability of approximately 0.38. Thus there is a a probability of 0.62 of being outside this region, and half as much, or 31%, associated with the region more than 0.5 standard deviations on the high side of the mean. Such a deviation from an exactly 50–50 split is not at all unlikely. So, the urn and marble problem has a clear answer: we cannot be confident that there are more brown than green marbles in the urn as a whole. Insofar as the real problem can be described by the model – which would depend on whether the persons polled adequately represent the whole population we are interested in, and on whether or not something like a sudden plunge in the price of gasoline turned 'greens' overnight into 'browns' – the survey is consistent with the hypothesis that large cars are not preferred to small.

It is worth emphasizing a point made more than once already: a larger sample with the same percentage deviation from an even split would be evidence for a different conclusion. For example, if the sample were 100 times larger, the same fraction on one side of the issue would correspond not to 0.5 standard deviations but to 5, on the assumption that there is no preference. Since a fluctuation of 5 standard deviations is *very* unlikely, it would then make sense to modify the assumption and conclude that there was indeed a small but significant preference.

Testing a vaccine

A statistical experiment involving 100 000 people, which was contemplated in the last paragraph, would be a large undertaking – but small compared to one precedent. In the summer of 1954 there was a test of the efficacy of the Salk polio vaccine involving over a million children. A study of the details of this enormous effort would raise practical questions far more subtle than those touched on in the last example, and moral ones as well: all I want to do here is to show that you already understand the reason underlying one part of this experiment. Approximately 400 000 children were randomly divided into two groups. One group was treated with the vaccine; the other received a placebo injection. Why was such a large number of subjects needed? The hypothesis to be tested was that the treatment approximately halved the chance of catching a serious form of the disease. Previous experience suggested that without any preventative measures the chance of a serious attack of polio in the course of a summer was about 5 in 10 000 for the age group (second graders) being treated. Assuming this, and that the attacks are independent random events, what range in the number of cases is 95% probable, for the vaccinated group and for the control group? The answer, as you already know, is: Two standard deviations on either side of the mean. Written algebraically, in terms of formulas worked out in the last chapter, and used in the last section, this yields the range

$$Np \pm 2\sqrt{Np(1-p)} \tag{4.2}$$

(In the present case, p is so small that no error is made by ignoring the factor $1-p$ within the squareroot.)

In Table 4.1, I have worked out a table of values obtained from this expression. One learns from the table that only group sizes of 100 000 or larger offer reasonable assurance that the number of cases among vaccinated individuals will be smaller than the number in the control group. For smaller samples, statistical fluctuations could disguise the protective effects, if any, of the vaccine.

The assumption most open to question in the calculations just done is that the attacks are random. Polio did in fact occur in epidemic waves. One can argue that the assumption is not too bad because the individuals in the test were chosen at random, and because the sample was taken from the entire country, i.e. a region large compared to the particular towns or suburbs where epidemics might occur. But, I am getting uncomfortably close to the world of a practicing statistician, and so, apart from reminding you that the test of 1954 did conclude that the vaccine was an effective preventative with beneficial results for all, I shall now relapse into whimsy.

Estimation of need

When my older daughter was a student at Cornell and living in a dormitory, she once came home with two boiled lobsters. The powers that be had decided to liven up the year with a lobster dinner, had badly overestimated the capacity of the dining population, and were desperately giving away the considerable excess. Now, I have no way of knowing precisely how they got themselves into this predicament, but it does raise the interesting question of how one should estimate a need of this kind; and I can think of one interesting way in which one could end up with an overestimate.

Suppose I am a cook with a lot of experience preparing lobster meals for small groups of undergraduates, and suppose I know that I should buy 11 or 12 lobsters to be sure of satisfying 10 diners.

Table 4.1. *Calculated range in the number of polio cases to be expected with 95% probability in groups of various sizes based on two assumptions for the probability of attack: $p = 5 \times 10^{-4}$, with no protection, and $p' = 2.5 \times 10^{-4}$, for the vaccinated group.*

Size of group N	Range $p = 5 \times 10^{-4}$	$p' = 2.5 \times 10^{-4}$
10 000	5 ± 5	3 ± 3
50 000	25 ± 10	13 ± 7
100 000	50 ± 14	25 ± 10
500 000	250 ± 32	125 ± 22

Then, faced for the first time with the job of feeding 1000, I might order 1150 lobsters. And so might you have done, until you started to read this book. Now you are in a position to be wiser. It must be emphasized, however, that the estimate you make will depend heavily on assumptions based on the past experience of our hypothetical cook. Suppose, then, that you question me, the cook, and I tell you that I have found from time to time that 9 or, very rarely, 8 lobsters would be all that 10 undergraduates would consume, because there was the occasional one who didn't care for them. Then you might reasonably surmise that what I am telling you is that the average undergraduate will eat one lobster, and that the extra one or two that I order for a group of ten is to avoid disappointing the occasional passionate lover of lobsters. Aha, you say, the excess over one per person is to take care of statistical fluctuations: when the sample size is scaled up by a factor of a hundred, as in going from a group of 10 to one of 1000, the fluctuations will scale by a factor of the squareroot of a hundred, i.e. ten! So, you will order 1015 lobsters, or, if you want to be extraordinarily careful, 1030. The number I would have ordered, based on a misinterpretation of my very relevant experience, would certainly result in an oversupply of 100 or thereabouts.

Fanciful though this explanation may be, it does bring out a point worth making, which is that in such matters the answer depends, and often sensitively, on assumptions it is hard to be sure about. Garbage in, as they say, garbage out. On the other hand, lacking an understanding of statistical fluctuations one has the worse possibility: sense in, garbage out.

You may be finding that you are developing an intuitive understanding of statistical fluctuations. The error made by our hypothetical cook was to imagine that in the 1000 people to be satisfied there might be a large number of lobster lovers. What he or she overlooked was that, according to past experience, lobster lovers are on the average matched by lobster haters, leading to a clustering about the most likely situation of one lobster consumed per diner.

The method of reasoning described here is used in many situations where demand has to be estimated. For example, a telephone exchange never has enough lines to satisfy every customer at the same time. The aim is to satisfy the average demand and to have some extra lines available to handle fluctuations. Most of the time, the system works very well. On the other hand, an unusual happening, such as a power failure or large storm, will cause almost everyone to pick up their phones and will overload the exchange.

Probability and baseball: the World Series

Say it ain't so, Joe

Anonymous

If probability enters our daily lives through weather reports and polls, statistics in the everyday sense of a quantitative record of past events are part and parcel of sports, with baseball leading the way. The preoccupation with past performance reaches a peak around the time of the World Series, a series that has nothing to do with the world and is won by the team that first wins four games. The World Series games I have watched (on TV) have turned on minor, almost random, events – a few lapses from perfection. In fact, the teams have seemed so evenly matched that I could easily have imagined the result going the other way. What if we make the (heretical) assumption that winning or losing a game in the world series has the same odds as tossing a fair coin? The assumption is certainly open to question, but let's work out its consequences first.

Here is a quote from *The New York Times* in October 1981 when the series was tied at two games apiece:

> Nobody needed reminders of the urgency of today's game, but reminders were hard to escape. This was the 30th time in the 78 years of the series that the teams had been tied after four games. And 22 times the team that won the fifth game had gone on to win the series.

To try out our assumption, we have to answer the question: If a fair coin is tossed four times, what is the probability of two heads and two tails? $\binom{4}{2} \times (\frac{1}{2})^4$ you cry with one voice. Canceling out common factors, we get $3/8$. The historical 30 in 78 tries could hardly be closer! (In fact, if we want to be *very* learned we can note that the standard deviation associated with 78 tries of an experiment with probability $3/8$ is $\sqrt{78 \times (3/8) \times (5/8)} \approx 4$. So, statistical fluctuations would allow for a significantly larger discrepancy from the predicted mean of $78 \times (3/8) \approx 29$ than actually occurred.) The hypothesis that teams which reach the world series are evenly matched, and evenly matched anew at the start of each game, has done pretty well.

What about the other statistic referred to in the quoted column having to do with the probability of winning the series after leading 3 to 2? With 5 games played, no more than two games remain. There are four possible outcomes of two games. If the team behind at this point wins both of the last two, it wins the series. The three other outcomes deliver the series to the other team. Thus, on the assumption that all outcomes are equally likely, the team ahead after 5 games has a chance of 3/4 of taking the series. The record of 22 in 30 could not be closer to this fraction! (Note that we included the last game even though it would not take place if the team ahead after 5 games had won the sixth. Although the result of a last game could then not change the previously determined final consequence, so that such a game would not be played, it is convenient to include it in order to count outcomes consistently and straightforwardly.)

Buoyed by these successes, let's try a more elaborate test. What, again on the assumption that every game is an independent random experiment with an equal probability for a win or a loss, are the probabilities for 4, 5, 6, or 7 game world series? As in the last paragraph, we have to make a consistent apportionment of the total number of outcomes if all games are played. For 7 games there are $2^7 = 128$ possible outcomes. If a 4 game series is to occur, one or other of the teams must win all of the first 4 games, and the last 3 'phantom' games can go either way. The number of possibilities can be seen to be $\binom{4}{4} \times 2 \times 2^3$, as follows. The combinatorial factor is a fancy way of writing 1 as the number of ways of chosing 4 games out of the first 4, the 2 accounts for the fact that either one of the two teams could win these first 4 games, and the final 2^3 corresponds to the possible outcomes of the 'phantom' games which do not influence the issue. We have thus found that 16 of the 128 outcomes are to be assigned to 4 game world series. Continuing in the same way, the number of 5 game world series is $\binom{4}{3} \times 2 \times 2^2$. In this case, one or other of the teams must win 3 of the first 4 games, and the winner of these games must win the 5th, leaving 2 'phantom' games that can go either way. Multiplying out the factors gives 32 possibilities. For

Table 4.2. *Lengths of World Series in two 25 year periods compared with mean numbers predicted on the assumption of equal likelihoods of winning or losing in each game.*

	1926-50	Calculation	1951-75
7 game series	7	7.8	15
6 game series	5	7.8	3
5 game series	7	6.2	4
4 game series	6	3.1	3

6 game series, you will verify the formula $\binom{5}{3} \times 2 \times 2$, which gives 40 cases. Proceeding similarly for 7 game series, we get $\binom{6}{3} \times 2$, which turns out also to be equal to 40. As a verification of the counting we note that $16+32+40+40=128$.

To compare these calculations with experience, I have listed in Table 4.2 the numbers of world series of various lengths for two 25 year periods, 1926–50 and 1951–75. In the middle column are the mean numbers for 25 occurrences as given by the probabilities calculated in the last paragraph. (For example, for 7 game series we have the predicted mean number $\frac{40}{128} \times 25 = 7.8$. And so on for the other entries in the middle column.) The striking thing about the table is that all the facts agree quite well with the calculations except for a glaring discrepancy in the number of 7 game series in the interval 1951 to 1975.

What are we to make of the 15 that sits in the first row of the last column? All that can be said is that it is *extremely* unlikely on the basis of the assumptions we have been making. Again, being learned, we can work out the standard deviation for an experiment with probability $40/128$ tried 25 times. The answer is $\sqrt{25 \times (40/128) \times (88/128)}$ which is 2.3. Two standard deviations on either side of 7.8 gives, to the nearest integer, the range 3–12. The budding statistician in us cries out that the hypothesis must be wrong.† But why does it work so well in the other cases? And, how should we modify it? It won't do to say that the teams were unequally matched, because that would give fewer long series. It could be that during these years the team that fell behind redoubled its efforts and equalized the series. So, ultimately, we can say nothing definitive. In a lottery or some other situation where the odds are independently known such an enormous discrepancy would indicate fraud.

This brings us to the end of our brief, partial, but intense study of the mathematical structure and applications outside science of the theory of probability. If the examples have suggested that such applications can be an uncertain business, that may have been my intention. But it was also my intention to illustrate that, when used with common sense, an understanding of the workings of chance gives deep insights, allows surprising predictions, and provides a framework for critical thought.

---

† For what it's worth, the number (6) of 7–game series in the 18 years 1976–93 falls well within the *one* standard deviation range, 5.6 ± 2.0, predicted using our hypothesis. (This is written in 1994, the year of the strike.)

Solved problems

(1) A drunk starts from a lamppost half way down a street taking steps of equal length L to the right or to the left with equal probability. [This process is called a 'random walk' in the theory of probability.] How far (measured in sober steps of length L) is he likely to be from the lamppost after taking N steps, where N is a large number? Explain

Solution: The repeated random experiment here is a step of length L to the right or the left with equal probability. Using (3.2) or (3.4), the mean distance moved per step, call it μ_1, is $L \times \frac{1}{2} + (-L) \times \frac{1}{2} = 0$. Using (3.3) or (3.5), the variance for a single trial, call it σ_1^2, is $L^2 \times \frac{1}{2} + (-L)^2 \times \frac{1}{2} = L^2$. After a large number, N, of steps, the probability of various numbers of steps to the right or left will approach a normal distribution with mean $\mu_N = N \times \mu_1 = 0$, and standard deviation $\sigma_N = \sqrt{N} \times \sigma_1 = \sqrt{N} \times L$. Now we are contemplating not many such drunken walks of N steps, but just one, so a criterion of likelihood is needed. For example, Table 3.5 on p. 42 informs us that it is 95% likely that the drunk will be found within a distance $2L\sqrt{N}$ to the right or left of the lamppost.

(2) A certain light-seeking bacterium swims in spurts of length ℓ. When confined to a narrow tube in an appropriate liquid, it swims in random spurts to the right and left but is twice as likely to move towards light as away from it.

A large number of bacteria are introduced into the middle of the tube. Each bacterium has enough room to swim independently of the others. Describe quantitatively the main features, and make a rough sketch, of the distribution of bacteria along the tube after a time such that each one has swum N spurts. N is a large number, but still such that $N\ell$ is less than half the length of the tube.

Solution: This problem is very similar to the previous one. Now, because of the greater probability of a spurt to the left on a single trial, the mean distance moved per step is $\mu_1 = \ell \times \frac{2}{3} + (-\ell) \times \frac{1}{3} = \frac{\ell}{3}$. [With reference to the drawing, where the light is to the left, we have chosen to measure distances positive to the left and negative to the right.] Now, the single trial variance may be calculated in two equivalent ways. We may either, as in (3.5), average the squares of the differences between the outcomes and the mean, or, as in (3.17), subtract the square of the mean from the average of the squares of the outcomes. It may be instructive to do this both ways. According to the first method:

$$\begin{aligned}
\sigma_1^2 &= (\ell - \frac{\ell}{3})^2 \times \frac{2}{3} + (-\ell - \frac{\ell}{3})^2 \times \frac{1}{3} \\
&= \ell^2 (\frac{8}{27} + \frac{16}{27}) = \frac{8}{9}\ell^2.
\end{aligned}$$

(4.3)

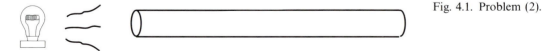

Fig. 4.1. Problem (2).

The second method gives:

$$\begin{aligned}
\sigma_1^2 &= \ell^2 \times \frac{2}{3} + (-\ell)^2 \times \frac{1}{3} - (\frac{\ell}{3})^2 \\
&= \ell^2(1 - \frac{1}{9}) = \frac{8}{9}\ell^2.
\end{aligned} \tag{4.4}$$

Both methods give the same answer, as they must, but the algebra is easier the second way. The mean and variance for N trials are N multiplied by the single trial mean and variance, respectively. Since we have many, many bacteria their distribution in space will closely approximate a normal distribution with mean $\mu_N = N \times (\ell/3)$ and standard deviation $\sigma_N = (2\ell/3)\sqrt{2N}$. The requisite sketch is given below.

(3) In a certain country 0.1 of the families have no children, 0.25 have one child, 0.5 have two children, and 0.15 have three children.

   (a) Calculate the mean and standard deviation of the number of children per family.

   (b) If 10 000 families are picked at random what is a likely range in the number of children one would find?

   Solution: The mean number of children per family is

$$\mu_1 = (0 \times 0.1 + 1 \times 0.25 + 2 \times 0.5 + 3 \times 0.15) = 1.7. \tag{4.5}$$

The variance, using the method discussed on p. 42 – which problem (2) will have convinced you, if you needed convincing, is the easier tactic – is:

$$\begin{aligned}
\sigma_1^2 &= [(0^2) \times (0.1) + (1^2) \times 0.25 + (2^2) \times 0.5 + (3^2) \times 0.15] - (1.7)^2 \\
&= 3.6 - 2.9 = 0.7
\end{aligned} \tag{4.6}$$

The standard deviation is thus $\sqrt{0.7} \approx 0.85$. In a group of 10 000 families picked at random it is then 95% likely, given the assumptions,

Fig. 4.2. Sketch graph for solution to problem (2).

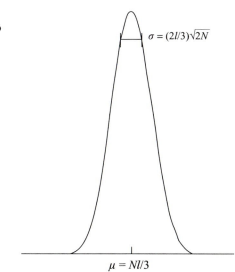

$\sigma = (2l/3)\sqrt{2N}$

$\mu = Nl/3$

that there will be a range of $10000 \times 1.7 \pm 2 \times 100 \times 0.85$, i.e 17000 ± 190 children. [As always, the mean per family must be scaled up by the number of families, whereas the standard deviation per family must be multiplied by only the squareroot of the number of families.] The small variability shows the law of large numbers at work.

(4) In a certain large city 1% of concertgoers require wheel-chair accessible seating. You are designing a concert hall to hold 2500 listeners. How many wheel-chair accessible seats would you install to be

(a) 50% certain of adequate seating for the disabled?

(b) 97.5% certain of adequate seating for the disabled?

[Note: 95% of the area of the normal distribution is contained in two standard deviations on either side of the mean.]

Solution: The easy part here is the calculation of the mean number of seats needed ($\mu = 2500 \times 0.01 = 25$), and the standard deviation in the number of seats needed ($\sigma = \sqrt{2500 \times 0.01 \times 0.99} \approx 5$. Careful thought is needed to relate these numbers to the questions (a) and (b). Glancing at Fig. 3.3, which includes a drawing of the normal distribution, we see that 50% of the probability occurs below the mean. Thus, if 25 seats are provided for the disabled there is an equal likelihood of more seats or less seats being required at a particular concert. This, then, is the answer to part (a). For part (b), note that since 95% of the probability is contained in two standard deviations on either side of the mean, 2.5% is contained in each of the wings. The condition posed in (b) requires that more seats be required on average only at 2.5% of concerts. These undesirable occurrences occur only at the high end of the distribution. Thus, two standard deviations more than the mean, or $25 + 2 \times 5 = 35$ seats, is the answer.

(5) The food store in Copenhagen airport, through which literally thousands of people pass each week, sells an average of 36 small jars of caviar per week. The shelves are replenished once a week. As a statistically literate and prudent manager, you wish to be 97.5% certain of having an adequate supply. How many jars would you have on hand at the start of a new week?

Solution: We have here, by hypothesis, the event of a person taking a jar of caviar with a very small probability, p, or choosing not to take one with a probability $1 - p$, very close to 1, which random experiment is repeated a very large number, N, times every week. In this respect the problem is very much like the last one. However, we are given neither N nor p but only the mean number of jars taken each week Np, which is given to us as 36. Interestingly, the standard deviation is determinable from this information. In general, for a binomial process we know it to be $\sigma = \sqrt{Np(1-p)}$. Now, because p by hypothesis is extremely small, this is very well approximated by $\sigma = \sqrt{Np} = \sqrt{36} = 6$. Our manager wants to have so many jars on hand at the start of each week that *more* would be required only 2.5% of the time. From the argument given in the solution to the last problem, it follows that 2 standard deviations on the high side of the mean achieves this aim. The required number of jars at the start of each week is thus $36 + 2 \times 6 = 48$. The answer is sensitive to the assumptions. Presumably, different circumstances prevail during holiday seasons.

5

A little mathematics

If a thing is worth doing, it is worth doing badly.
                                        G.K. Chesterton†

The only mathematical operations we have needed so far have been
addition, subtraction, division, multiplication, and an extension of the
last: the operation of taking the square-root. [The square-root of a
number *multiplied* by itself is equal to the number.] We have also had
the help of a friendly work-horse, the computer, which gave insight
into formulas we produced. That insight was always particular to the
formula being evaluated; the general rules of thumb we used in the
last chapter were obtained by nothing more or less than sleight of
hand. To do better, there is no way of avoiding the mathematical
operation associated with taking arbitrarily small steps an arbitrarily
large number of times. This is the essential ingredient of what is called
'calculus' or 'analysis.' But don't be alarmed: our needs are modest.
We shall be able to manage very well with only some information
about the exponential and the logarithm. You will have heard, at least
vaguely, of these 'functions' – to use the mathematical expression for a
number that depends on another number, so that it can be represented
by a graph – but rather than relying on uncertain prior knowledge,
we shall learn what is needed by empirical self discovery.

Powers

The constructions we need are very natural generalizations of the
concept of 'powers' which you will find that you are quite familiar

---

† The witticism is also wise, provided one sees through some of its shine by properly interpreting
the word badly – as something like: in a less than completely thorough way. Then it becomes a
lapidary statement of one of my motives for writing this book. I take it as my text because the time
has come to do something I can here only do 'badly,' which is to introduce, and demonstrate the
utility of, certain simple constructs that occur so often in useful mathematics that they have been
studied, tabulated, and given names.

with from what has gone before. When we multiplied a number by itself, or squared it, we wrote it as the number raised to the second power, i.e., if x was the number we wrote $x \times x = x^2$. The rules for combining positive integer powers n and m are

$$x^n \times x^m = x^{n+m}, \tag{5.1}$$

$$(x^n)^m = x^{nm}. \tag{5.2}$$

These equations are proved by saying in words what the left-hand sides mean, and finding that one thereby articulates the meanings of the right-hand sides. Try it with (5.1): If n factors of x are further multiplied by m factors of x, the answer is $n+m$ factors of x. Similarly, (5.2) says that if the quantity n factors of x is multiplied by itself m times, the result is nm factors of x.

Equation (5.1) is consistent with the convention:

$$x^0 = 1. \tag{5.3}$$

because 1 is the number which does not change any number it is multiplied by.

Fig. 5.1. Graphs of x^n for various values of n

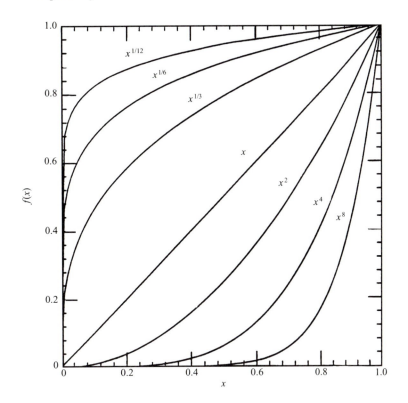

By requiring that the rules (5.1)–(5.2) hold more generally, one can also give meaning to negative and fractional 'powers' of a positive number x. From (5.1) and (5.3), one obtains the interpretation $x^{-n} = 1/x^n$. (Stop and check that there is no other way of making (5.1) compatible with (5.3) for $m = -n$.) Similarly, from (5.2) it follows that $x^{1/n}$ must be interpreted as $\sqrt[n]{x}$, the nth root of x, i.e. the positive number which, when multiplied by itself n times, gives x. (Verify that this is the only way of having (5.2) be true for m=(1/n).) Fractional powers $x^{m/n}$ must then be interpreted as $\sqrt[n]{x^m}$.† An ordinary number can be approximated to any desired accuracy by a fraction; so we have, in principle, a way of constructing the function x^y for x non-negative and y any ordinary number. Because of its usefulness, a procedure for calculating this function is built into pocket calculators for scientific applications. Figure 5.1 is a graph of this function of x for several values of y.

Exponentials

The exponential is our friend from the previous paragraphs, x^y, now, however, considered as a function of y for fixed x. In order to keep to the usual convention of x being the independent variable, we shall reverse the roles of x and y and speak about the 'function of x' given by y^x, for fixed y.

Most frequently, one encounters this function for the case $y = e$, where e is the number defined below which can be written to any desired accuracy as a decimal, and which to 9 decimal places is given by 2.718281828. This remarkable number occurs naturally in many contexts. Compound interest is one such example.

Consider an interval of time at the end of which simple interest at 100% is paid. Then $1 invested returns $1 in interest, or a total of $2. If the interest is compounded halfway through the period, the return is $ $(1 + \frac{1}{2})$ at the end of half the period; this sum reinvested for the remainder of the period yields a total return of $1\frac{1}{2}$ times the sum reinvested, or $ $(1 + \frac{1}{2})^2 = \$2.25$. Imagine subdividing the interval more and more finely, and reinvesting the proceeds at the end of each subdivision. For an N-fold division the total yield would be $ $(1 + \frac{1}{N})^N$. Is this a finite number for very large N? Well, it is. With the help of

---

† What at first sight may seem like an alternate interpretation, $(\sqrt[n]{x})^m$, is another way of writing the same thing, as can be seen by raising both expressions to the nth power to get x^m in each case.

my trusty calculator, and the operation y^x, I discover:

$$(1 + \frac{1}{10})^{10} = 2.593742460\ldots$$

$$(1 + \frac{1}{100})^{100} = 2.704813829\ldots$$

$$(1 + \frac{1}{10^6})^{10^6} = 2.718280469\ldots$$

$$(1 + \frac{1}{10^9})^{10^9} = 2.718281827\ldots$$

It would seem, and it is true, that as N gets larger the operation described above converges to a definite number. This number is called e, and one writes the limiting expression for it as

$$e = \lim_{N\to\infty} (1 + \frac{1}{N})^N. \tag{5.4}$$

This is read as 'e equals the limit as N goes to infinity of ...'

Now what about a situation in which the annual interest rate is 5%? Simple interest on an investment of \$1 yields \$1.05. Compounded twice, the yield is $(1 + \frac{0.05}{2})^2 = 1.050625$, which is still \$1.05. Compounded daily, the yield is $(1 + \frac{0.05}{365})^{365} = 1.0512\ldots$, i.e. still \$1.05. Rapid compounding does not add very much to a low interest rate, as the banks well know. What about the limiting case:

$$\lim_{N\to\infty} (1 + \frac{0.05}{N})^N = \lim_{M\to\infty} (1 + \frac{1}{M})^{0.05M} = \lim_{M\to\infty} \left[(1 + \frac{1}{M})^M\right]^{0.05}$$
$$= e^{0.05} = 1.051271096. \tag{5.5}$$

The second step is justified: you cannot stop me from replacing the large number N by the large number $M = (N/0.05) = 20N$.

Here, then, is a simple example that leads one to be interested in the function e^x for x a real positive number. From the above work, we have

$$e^x = \lim_{N\to\infty} (1 + \frac{x}{N})^N = \lim_{M\to\infty} (1 + \frac{1}{M})^{Mx}. \tag{5.6}$$

From the rule for powers (5.2), and the finiteness of the limit in (5.4), we see that x does indeed play the role of a power. The function e^x is plotted in Fig. 5.2. It grows very rapidly as x is increased. This is 'exponential' growth, characterized, as our example of compound interest has shown, by an increase in every small interval proportional to the value at the beginning of the interval. It is natural that the function should occur in the growth of idealized populations and economies. Since x occurs as a power, a formula for $(1/e^x)$ is obtained from (5.6) by changing x to $-x$, i.e. $(1/e^x) = e^{-x}$. Note that e^x grows with increasing x because e is a number greater than 1. The function y^x for y fixed and less than unity falls for increasing x. This is illustrated in Fig. 5.2 for $y = \frac{1}{2}$.

In our later work, we shall find it useful to be able to write e^x as a sum of integer powers of x. Such an 'expansion' can be obtained from (5.6) and a knowledge of how to express $(1 + a)^N$ as a sum of powers of a. For $N = 2$, I have assumed, e.g. above (3.12), that you know the answer $1 + 2a + a^2$. In fact you can put together the general case from things you have already learned. Since $(1 + a)^N$ is N factors of $1 + a$ multiplied together, it will contain all integral powers of a from 0 to N. The coefficient of a^r is the number of ways in which r

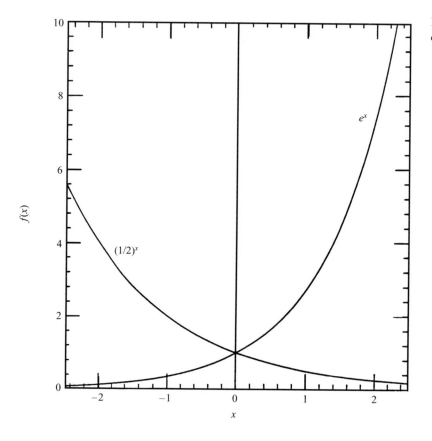

Fig. 5.2. Graphs of e^x and $(\frac{1}{2})^x$.

factors of a can be chosen from the N available factors. But this is $\binom{N}{r}$. Thus,

$$
\begin{aligned}
(1 + a)^N = 1 + Na &+ \frac{N(N-1)}{2!} a^2 \\
&+ \frac{N(N-1)(N-2)}{3!} a^3 + \cdots,
\end{aligned}
\tag{5.7}
$$

where the dots signify terms proportional to a^r with r running from 4 to N, and the coefficients can if you wish be read from the $(N + 1)$st row of Pascal's triangle on p. 43.

Using (5.7) in the second form in (5.6) yields

$$
\begin{aligned}
e^x = \lim_{N \to \infty} (1 + \frac{x}{N})^N = \lim_{N \to \infty} \Bigg[1 + N(\frac{x}{N}) \\
+ \frac{N(N-1)}{2!} (\frac{x}{N})^2 + \frac{N(N-1)(N-2)}{3!} (\frac{x}{N})^2 + \cdots \Bigg].
\end{aligned}
\tag{5.8}
$$

Within the large square bracket there are $(N+1)$ terms, but we have somehow to take the limit as N becomes arbitrarily large. This can be achieved by dividing out the factors of N that go with each power of x to obtain

$$
e^x = \lim_{N \to \infty} \Bigg[1 + x + \frac{1}{2!} (1 - \frac{1}{N})x^2 + \frac{1}{3!} (1 - \frac{1}{N})(1 - \frac{2}{N})x^3 + \cdots \Bigg]
\tag{5.9}
$$

As N is allowed to become arbitrarily large, terms of the form $\frac{1}{N}, \frac{2}{N}$, etc. can be replaced by zero, so that one has achieved the 'infinite series expansion'

$$
e^x = 1 + x + \frac{x^2}{2!} + \frac{x^3}{3!} + \frac{x^4}{4!} + \cdots
\tag{5.10}
$$

Infinite series have to be treated with some caution, since it is easy to construct examples whose sum is infinite. Because of the factorials in the denominators, (5.10) is a particularly robust series which can be used for any finite numerical value of x.† For very small x it is clear that the terms are successively smaller. This fact allows one insight into the neighborhood of any point on the curve of e^x shown in Fig. 5.2. Let h be a number much less than 1. Then

$$
e^{x+h} = e^x e^h \approx e^x [1 + h].
\tag{5.11}
$$

[The symbol $\approx$ was already encountered in the problems at the end of Chapter 3; it means and is said 'approximately equal to.'] In (5.11), the second form is identical – using the property (5.1) of powers – to

---

† For an example of what this infinite series is good for, see Problem 5 at the end of this chapter.

the first, and the last is an approximation – good for very small h – that includes only the first two terms of (5.10). Equation (5.11) shows that when the independent variable x is increased by a small amount, the function e^x changes by an amount proportional to its numerical value at x. This implies that, since e^x increases rapidly with x as shown in Fig. 5.2, the *change* of this function for a small constant change of x increases equally rapidly, a fact that is not quite so obvious from the figure.

Logarithms

Now, a few words about logarithms. These are closely related to exponentials. If y and X are positive numbers, and we can find a number x such that

$$y^x = X \tag{5.12}$$

we say that $x = \log_y X$. In words, x is the logarithm of X to the base y. A common base is 10. Thus, $10^2 = 100$ and $\log_{10} 100 = 2$ are equivalent statements.

From the properties of powers, it follows that:

$$\log(XZ) = \log X + \log Z \tag{5.13}$$

$$\log(X^s) = s \log X \tag{5.14}$$

$$\log 1 = 0. \tag{5.15}$$

These equations are true for any (positive) base. Equation (5.13–14) show that logarithms have the useful property of turning multiplication into addition.

Since e^x is a commonly occurring function, logarithms to the base e, called natural logarithms and written ln, are also common. The natural logarithms have a simple series expansion that is often useful. Let $x = \ln s$. Then $e^x = s$. As x approaches 0, s approaches 1, as is true for logarithms to any base. The special thing about natural logarithms is the behavior near $s = 1$. Write $s = 1 + t$. Then, using (5.10)

$$e^x = 1 + x + \frac{x^2}{2!} + \frac{x^3}{3!} + \cdots = 1 + t. \tag{5.16}$$

Equation (5.16) shows how t depends on x when x is small. We can ask the inverse question: How does x depend on t when t is small? Assume $x = a_1 t + a_2 t^2 + a_3 t^3 \cdots$, where a_1 a_2, etc. are unknown coefficients. Substituting this form into (5.13), and equating coefficients of equal powers of t, one finds $a_1 = 1$, $a_2 = -\frac{1}{2}$, and $a_3 = \frac{1}{3}$. We thus have the expansion:

$$\ln(1 + t) = t - \frac{t^2}{2} + \frac{t^3}{3} - \cdots \tag{5.17}$$

Note that in contrast to (5.10) there are no factorials in this expression. It is also valid in a more restricted range. It can be added up to give a sensible answer only when t is greater than -1 and less than or equal to $+1$.

In Fig. 5.3, $\log_{10} x$ and $\ln x$ are plotted. A good way to check that you understand the intimate connection between logarithms and exponentials is to explain to yourself how the plot in Fig. 5.2 of e^x is related to the curve in Fig. 5.3 of $\ln x$.

Fig. 5.3. Graphs of $\log_{10} x$ and $\ln x$

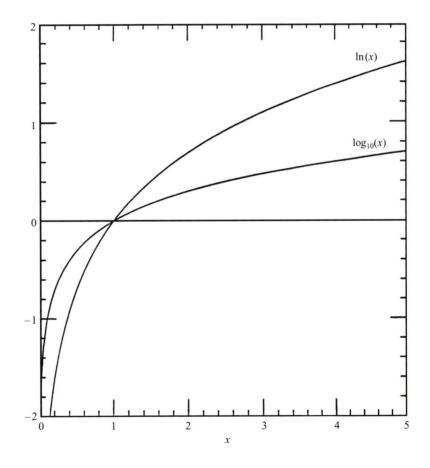

Radioactive decay

From now on we shall be hard put to do without the exponential
and the logarithm, which have many, many uses. An example from
physics, and one that also involves probabilities, concerns the decay
of unstable particles, such as radioactive nuclei or the sub-atomic
particles called mesons. Our best present understanding is that one
can not predict exactly when an unstable particle will change into
its decay products (other particles.) The theory we have for dealing
with this phenomenon, Quantum Mechanics, has the bizarre feature
of only predicting probabilities even when the starting conditions are
specified as precisely as possible.† This theory, which the most careful
measurements confirm, says that if the same decay process is observed
repeatedly it will occur at different times; the precise statement it
makes is that the *probability* of decay in any small interval of time is
proportional to that interval. Divide a given time t into M intervals,
where M is chosen sufficiently large that t/M is 'small' in the sense
mentioned above and to be made more precise shortly. The probability,
p, of decay in the interval (t/M) is then

$$p = \Gamma \frac{t}{M}, \tag{5.18}$$

where we have called the coefficient of proportionality Γ (capital
'gamma'). The probability of *no* decay in the small interval is thus

$$1 - p = 1 - \Gamma \frac{t}{M}. \tag{5.19}$$

Let us now calculate the probability, call it $\bar{P}(t)$, that a certain unstable
particle has *not* decayed in a time t. This probability is the product of
the (independent) probabilities of no decay in each of the M intervals
of duration t/M, by which chain of reasoning we find

$$\bar{P}(t) = (1 - \frac{\Gamma t}{M})^M \rightarrow e^{-\Gamma t}. \tag{5.20}$$

In the last step we have let M get arbitrarily large, indicated that
process by the symbol $\rightarrow$, and used (5.8) with x replaced by $-\Gamma t$.
From (5.20) one would conclude that the probability of no decay in a
time t/M is $e^{-\Gamma(t/M)}$, in apparent contradiction to (5.19). However, the
series expansion (5.10) shows that our starting point (5.18) is consistent
if (t/M) obeys the condition $\Gamma(t/M) \ll 1$, where the symbol $\ll$ means
'much less than.' This condition thus gives meaning to the word 'small'

† This matter is discussed further in Chapter 12

as used above. The quantity Γ is called the decay rate, and Quantum Mechanics gives a prescription for calculating it.

Another way of expressing the idea of 'exponential decay,' expressed by (5.20), may be familiar to you from discussions of the waste products of nuclear power plants – the concept of 'half-life.' If at time $t = 0$ there are $N(0)$ identical unstable particles each of which decays according to the exponential rule – with the same Γ, since the particles are identical – then on the average the number of such particles remaining after time t will, using (4.1), be

$$N(t) = N(0)e^{-\Gamma t}. \tag{5.21}$$

At what time will $N(t)$ be $\frac{1}{2}N(0)$? Let us call this time, the 'half-life,' T. Then, by assumption

$$N(T) = N(0)e^{-\Gamma T} = \tfrac{1}{2}N(0). \tag{5.22}$$

From this it follows that

$$e^{-\Gamma T} = \frac{1}{2}. \tag{5.23}$$

Now, one may eliminate Γ in favor of T in (5.21) because $e^{-\Gamma t} = (e^{-\Gamma T})^{(t/T)}$. Thus, using (5.23) in (5.21), we see that

$$N(t) = N(0)(\frac{1}{2})^{(t/T)}. \tag{5.24}$$

The function $(\frac{1}{2})^x$ is plotted in Fig. 5.2. You will verify from this plot that in $\frac{3}{4}$ of the half-life roughly 60% of a collection of identical unstable particles have not decayed.

As an exercise in the use of logarithms, let us work out how the decay rate and the half-life are numerically related. From the connection (5.23), it follows by taking the reciprocal of both sides that $e^{\Gamma T} = 2$. Now take the logarithm to the base e of both sides of this last equation to get

$$\Gamma T = \ln 2 = 0.6931 \tag{5.25}$$

where the final number is obtained either from a calculator or from a table of natural logarithms.

Note that the half-life is a time, whereas Γ is the probability of decay per unit of time, and is thus measured in the reciprocal of – i.e. unity divided by – units of time. Consequently, ΓT is a pure number, independent of whether time is measured in seconds or years, requiring only that the same unit be used in both factors on the left hand side of (5.25).

Logarithms and factorials: Stirling's formula

Both for the purposes of the next section – which is concerned with the startling agreement between the binomial distribution and the 'bell-shaped curve' or 'normal' distribution – and for the physical applications to come in later chapters, we shall need a mathematically tractable approximation for the factorials of large integers. This is provided by Stirling's formula:

$$r! \sim \sqrt{2\pi r}\, (r/e)^r. \tag{5.26}$$

The symbol $\sim$ means that the left hand side is well approximated by the right when r is large. To see how well this formula does, consider Table 5.1 in which the logarithms of both sides of (5.26) are compared. [Note the use of (5.13–14) in working out the formula for the logarithm of the right-hand side of (5.26).]

Here is an elementary derivation of an inequality that contains some of the essential terms of (5.26). The following identities are self-evident:

$$1! = 1 = 1,$$

$$2! = (1/2) \times 2^2 = \frac{1}{(1+1)} \times 2^2, \tag{5.27}$$

$$3! = (1/2) \times (2/3)^2 \times 3^3 = \frac{1}{(1+1) \times (1+\frac{1}{2})^2} \times 3^3.$$

For r an integer greater than 4, the above may be generalized as follows:

$$r! = \frac{1}{(1+1) \times (1+\frac{1}{2})^2 \times (1+\frac{1}{3})^3 \cdots \times (1+\frac{1}{r-1})^{r-1}} \times r^r. \tag{5.28}$$

All this has been without approximation. Now notice that the denominator of the fraction on the right of the last equation is the product

Table 5.1. *Comparison of* $\ln r!$ *with Stirling's approximation for it.*

r	$\ln r!$	$(r+\frac{1}{2})\ln r - r + \frac{1}{2}\ln 2\pi$
5	4.787	4.771
10	15.104	15.096
20	42.335	42.331
25	58.003	58.000
30	74.658	74.655

of $r - 1$ compound interest approximants to e, precisely as discussed above (5.4). Each of these approximants is *less than* e, as also shown above (5.4). The reciprocals of these approximants are therefore each greater than the reciprocal of e. Thus from (5.28) there follows the inequality

$$r! > \frac{r^r}{e^{r-1}}. \tag{5.29}$$

Since the logarithm illustrated in Fig. 5.3 is a steadily increasing function, the logarithm of the left side of (5.29) is greater than the logarithm of the right side, from which we see that $\ln r! > r \ln r - (r-1)$. As r gets larger and larger, more and more of the approximants to e in the denominator of (5.28) become close to e, so that the inequality (5.29) approaches an equality.

Stirling's formula is an only slightly better approximation to $r!$ than the one just worked out.

Near normality

Stirling's formula will enable us to show that the binomial distribution has a simple form near its maximum. In this way, we shall obtain the universal 'bell-shaped' curve. This calculation, in my experience, is as indigestible to some readers as it is soothing to others, so it is worth pointing out that it will not be directly used in the remainder of the book. It is here to ease the puzzlement of those who find the discussion in Chapter 3 too mysterious to be swallowed without explanation. Those who find the details that follow disturbing should ignore them: although each step is no more complicated than ones we have taken together, several are needed to complete the job.

To avoid irrelevant complications, we shall consider a binomial process with equally likely outcomes. Then from (3.1) the distribution corresponding to N trials is

$$P(r) = \frac{1}{2^N} \frac{N!}{(N - r)! r!}. \tag{5.30}$$

where r, the number of 'successes', runs from 0 to N. We already know that $P(r)$ changes rapidly near its single peak. Since the logarithm is a slowly varying function, it would seem like a useful tactic to consider the logarithm of (5.30) in an attempt to find a good approximation in the vicinity of the peak. Using the properties of logarithms (5.13–14), we find

$$\ln P(r) = \ln N! - \ln(N - r)! - \ln r! - N \ln 2. \qquad (5.31)$$

Now, suppose that N is large, and that r is somewhere in the neighborhood of $\frac{1}{2}N$ and thus also large. Each of the factorials in (5.30) is then well approximated by Stirling's formula (5.26). Using the expression for $\ln r!$ obtained from Stirling's formula and written out at the top of the third column of Table 5.1, one then finds, after collecting terms

$$\ln P(r) \sim (N + \tfrac{1}{2}) \ln N - (N - r + \tfrac{1}{2}) \ln(N - r)$$
$$- (r + \tfrac{1}{2}) \ln r - \tfrac{1}{2} \ln 2\pi - N \ln 2. \qquad (5.32)$$

In the region of the peak, r is near $N/2$. To emphasize this region of interest, let us introduce a new variable x defined by the relation $r = (N/2) + x$. Making this substitution in (5.32), you can verify that this becomes, without any further approximation

$$\ln P(\frac{N}{2} + x) \sim x[\ln(1 - \frac{2x}{N}) - \ln(1 + \frac{2x}{N})] - (\frac{N+1}{2})$$
$$[\ln(1 - \frac{2x}{N}) + \ln(1 + \frac{2x}{N})] - \tfrac{1}{2} \ln 2\pi N + \ln 2. \qquad (5.33)$$

Properties of logarithms have been used to obtain this form. A slightly tricky one is

$$\ln(\frac{N}{2} + x) = \ln[\frac{N}{2} \times (1 + \frac{2x}{N})]$$
$$= \ln N - \ln 2 + \ln(1 + \frac{2x}{N}). \qquad (5.34)$$

Equation (5.33) has been organized to show that x, the deviation of r from the mean $N/2$, only enters in the form $2x/N$. Now, the standard deviation of the distribution we are considering is $\frac{1}{2}\sqrt{N}$. In the region where x is positive or negative but of a magnitude less than several standard deviations, the ratio $2x/N$ is much less than unity when N is large. As a consequence, we may use the expansion (5.17) for each of the logarithms in (5.33). Furthermore, because N is much greater than 1, we may replace $N + 1$ by N in the third term on the right of this expression. One finds that the coefficients of the terms proportional to x and x^3 are zero, so that there emerges the approximate answer, valid for N large and $\frac{2x}{N} \ll 1$:

$$\ln P(n) \approx \ln(\frac{2}{\sqrt{2\pi N}}) - \frac{2x^2}{N}. \qquad (5.35)$$

To write this in a standard form, we undo the logarithm by taking the exponential of both sides. It is then also convenient to recognize $\frac{1}{2}N$ as the mean of our distribution (call it μ), and $\frac{1}{2}\sqrt{N}$ as the standard deviation (call it σ). These operations and substitutions lead to the simple result

$$P(n) \approx \frac{1}{\sigma\sqrt{2\pi}} \, e^{-\frac{(n-\mu)^2}{2\sigma^2}}. \tag{5.36}$$

This is the famous normal distribution with the properties listed in Table 3.5. The curve drawn in Fig. 3.3 is this function with $\mu = 40$ and $\sigma^2 = 20$.

The result obtained here is more general than the special case for which it was derived. It requires only a little more work, none of it beyond your reach, to show that the same form describes the binomial distribution with unequal probabilities, p and $1 - p$, for the two outcomes, in which case (4.1) applies so that $\mu = Np$ and $\sigma^2 = Np(1 - p)$. To show that the same form also emerges for many independent trials of a random event which is itself described by a distribution is more complicated. Provided the distribution for the single event is not too pathological, it turns out that the normal distribution again well describes the situation a few standard deviations on either side of the mean. As in the discussion leading to (3.9) and (3.11), the N trial mean and standard deviation which now enter (5.36) are related to the single trial mean and standard deviation by the formulas $\mu_N = N\mu_1$ and $\sigma_N = \sqrt{N}\sigma_1$.

This last section has not been easy. Although it has taken the mystification but I hope not the magic out of Chapter 3, it is not essential for anything that follows. The exponential, the logarithm, Stirling's formula, and a few other mathematical operations that we shall develop in the problems to follow, will give us the tools needed to explore the ways in which probability has entered the world of physics.

Appendix: The Poisson distribution

As another example of a simple situation in which the exponential function occurs in the context of probability, consider what happens to the binomial distribution (3.1) when the number of trials, which we have always called N, becomes extremely large, and the success probability per trial, which we have been calling p, becomes very small, but in such a way that the product Np remains finite. [Specific examples will be given below.]

Since this is a special case of the binomial distribution, the mean number of successes is given by our much quoted formula, $\mu = Np$. The limiting process referred to in the last paragraph does, however, change the formula for the standard deviation. The exact expression (4.1) is $\sigma = \sqrt{Np(1-p)}$. Because p is very small, we can ignore it in the last bracket under the square-root, obtaining

$$\sigma = \sqrt{Np} = \sqrt{\mu}. \tag{5.37}$$

Now we can ask what happens to the whole distribution in the limit. In general we had the expression given in (3.1):

$$P(r) = \binom{N}{r} p^r (1-p)^{N-r}. \tag{3.1}$$

This can be manipulated without approximation as follows:

$$P(r) = \frac{(Np)^r}{r!} (1 - \frac{pN}{N})^N \times \frac{N \times (N-1) \times \ldots (N-r+1)}{N^r} \times (1-p)^{-r}. \tag{5.38}$$

Now, everything after the first multiplication sign on the right side goes in the abovementioned limit to 1. Replacing Np by μ in the the first two terms, and noting the occurrence of the form given in (5.8) we get what is called the Poisson distribution

$$P(r) = \frac{\mu^r}{r!} e^{-\mu}. \tag{5.39}$$

The Poisson distribution has the feature that the whole distribution is determined by the mean. We already encountered an example of its applicability in Problem 5 at the end Chapter 4 where we estimated the need for jars of caviar in our (fictitious) shop in Copenhagen airport. In that problem, however, we were only interested in the standard deviation, which we worked out to be $\sqrt{\mu}$ as in (5.37).

A beautiful, and much quoted, example of a Poisson distribution occurs in some records kept by the German army between 1875 and

Table 5.2.

Deaths per year	Number of corps
0	144
1	91
2	32
3	11
4	2
5 or more	0

1894. This army had a very large number of soldiers grouped into corps, and a very small number of deaths caused by the kicks of horses. The data is given in Table 5.2.

The total number of corps in the sample you will verify to be 280. The recorded mean number of deaths per corps per year is:

$$\frac{1}{280}[0 \times 144 + 1 \times 91 + 2 \times 32 + 3 \times 11 + 4 \times 2] \tag{5.40}$$
$$= 0.70.$$

Taking this number as the mean of a Poisson distribution, i.e., using the formula

$$P(r) = \frac{(0.7)^r}{r!}e^{-0.7} \tag{5.41}$$

one can easily work out the theoretical probabilities for various values of r. These are given in the second column of Table 5.3. The final column gives the observed occurrences as fractions, i.e., the entries in the second column in Table 5.2 divided by 280. There is a startling agreement between the two last columns. Thus the hypothesis that the deaths of soldiers by the kicks of horses during the period in question was a rare random event and governed by the rules that apply to such processes is vindicated. The input of *one* number – the mean calculated from the observations in (5.40) – and an understanding of the logic that underlies chance has given insight into the *six* numbers in the right hand column of Table 5.3.

Evidently it is possible to reason about bad as well as good luck.

Solved problems

(1) Prove that formula (5.7) works not only when the power is an integer but also when it is an ordinary number, x, positive, negative, or zero. In

Table 5.3.

n	$P(n)$	Observed fraction
0	0.50	0.51
1	0.35	0.33
2	0.12	0.11
3	0.03	0.04
4	0.005	0.01
5	0.0007	0

short, prove that for $a \ll 1$

$$(1+a)^x = 1 + xa + \frac{x(x-1)}{2}a^2 + \ldots.$$ (5.42)

Hint: Prove $(1+a)^x = e^{x\ln(1+a)}$, and then use (5.17) and (5.16).

Solution: Since for any positive number, z, $z = e^{\ln z}$, and since $\ln(1+a)^x = x\ln(1+a)$, the identity in the hint is true. Now, using (5.17) we have

$$x\ln(1+a) = xa - x\frac{a^2}{2} + \cdots,$$ (5.43)

whereupon (5.16) gives

$$e^{x\ln(1+a)} = e^{xa - \frac{xa^2}{2} + \cdots}$$

$$= 1 + [xa - \frac{xa^2}{2}] + \frac{1}{2!}[xa - \cdots]^2 + \cdots$$

$$= 1 + xa + (-x + x^2)\frac{a^2}{2} + \cdots$$ (5.44)

$$= 1 + xa + x(x-1)\frac{a^2}{2} + \cdots.$$

This is the desired form. Note that we have systematically neglected terms proportional to the third or higher powers of a, which would carry the expansion beyond what we were asked to prove. As a matter of fact, the next term shown in (5.7), and indeed higher terms in this expansion also apply to the case we have just discussed, but there are less cumbersome ways to show that.

(2) Derive the infinite series expansion

$$\frac{1}{1-a} = 1 + a + a^2 + a^3 + \ldots,$$ (5.45)

where a is a positive or negative number of magnitude less than 1.

This is a special case of the previous problem, but you may find it instructive to proceed as follows:

Define the finite series $S_N = 1 + a + a^2 + \cdots + a^N$, where the the dots mean a sum of all powers of a between 2 and N. Show that $aS_N + 1 = S_N + a^{N+1}$. Solve this equation for S_N, and then see what happens when N becomes arbitrarily large.

Solution: By explicit calculation, it is easy to show that $aS_N + 1$ and $S_N + a^{N+1}$ are each equal to

$$1 + a + a^2 + \cdots + a^{N+1}$$ (5.46)

so the proposed identity is indeed true. By re-arranging terms in this identity, it follows that $S_N(1-a) = (1-a^{N+1})$, so that, using the definition of S_N,

$$S_N = 1 + a + a^2 + \cdots + a^N = \frac{1 - a^{N+1}}{1 - a}.$$

Now, if a is smaller than 1, a^{N+1} approaches zero as N becomes large, so the numerator of the fraction on the extreme right approaches 1. Allowing N to become arbitrarily large thus proves the required identity.

(3) Prove by working out three terms that the series (5.10) is consistent with the multiplication rule for exponentials

$$e^{x+y} = e^x e^y. \tag{5.47}$$

Solution: Working as far as squared quantities we obtain for the right hand side:

$$[1 + x + \frac{x^2}{2} + \cdots][1 + y + \frac{y^2}{2} + \cdots]. \tag{5.48}$$

Now, multiplying each factor in the first bracket by every one on the right we get

$$1 + x + y + (\frac{x^2}{2} + xy + \frac{y^2}{2}) + [\frac{xy^2}{2} + \frac{x^2y}{2} + \frac{x^2y^2}{4}]. \tag{5.49}$$

Note that the three terms in the round brackets are equal to $\frac{1}{2}(x + y)^2$. The terms in the square bracket have higher powers than the second and it is not consistent to keep them when we have ignored the $x^3/3!$ and the $y^3/3!$ in (5.48). To quadratic terms (5.47) has thus been verified. [You may also verify that the cubic terms just mentioned when added to the first two in the square bracket in (5.49) combine to give $(x + y)^3/3!$, the term involving three powers of x and y, separately or in combination, on the left side of (5.47).]

(4) Prove by working out two terms that the series (5.17) is consistent with the rule for logarithms of products (5.13), which requires that

$$\ln[(1 + t)(1 + s)] = \ln(1 + t) + \ln(1 + s). \tag{5.50}$$

Solution: Multiplying out the product of factors in the square bracket on the left of (5.50) yields $[1 + t + s + ts]$. The use of (5.17) then gives:

$$\ln[1 + t + s + ts] = (t + s + ts) - \frac{1}{2}(t + s + ts)^2 + \cdots \tag{5.51}$$

We only want the result to terms containing two powers of t and s, separately or in combination. This means that we may neglect the ts in the last bracket of (5.51), because squaring that bracket would otherwise produce higher powers. With this neglect, (5.51) yields $t + s + ts - \frac{1}{2}t^2 - \frac{1}{2}s^2 - ts$. The last ts, which came from $(t + s)^2$, cancels the first, and one is left, after rearranging terms, with

$$t - \frac{1}{2}t^2 + s - \frac{1}{2}s^2, \tag{5.52}$$

which is equal to the right hand side of (5.50) to quadratic order.

(5) Show from the explicit form of the Poisson distribution (5.39) that the infinite series written below [using the shorthand for sums introduced in (3.4) but now generalized to include an infinite number of terms] can be added up to give results we could have deduced directly from the

Binomial distribution:

$$\sum_{r=0}^{\infty} P(r) = 1. \tag{5.53}$$

$$\sum_{r=0}^{\infty} rP(r) = \mu. \tag{5.54}$$

$$\sum_{r=0}^{\infty} (r - \mu)^2 P(r) = \sigma^2 = \mu. \tag{5.55}$$

Solution: Equation (5.53) constitutes an explicit verification that the Poisson distribution is normalized. The proof follows from the infinite series (5.10) for e^x, as follows. Substituting (5.39) into (5.53) and writing the sum explicitly gives

$$\sum_{r=0}^{\infty} P(r) = [1 + \mu + \frac{\mu^2}{2!} + \frac{\mu^3}{3!} + \cdots]e^{-\mu} = 1. \tag{5.56}$$

Here the dots indicate the continuation of the series, which (5.10) tells us is nothing else but e^{μ}. The last step then follows from $e^{\mu}e^{-\mu} = 1$, and the 'normalization' is proved.

To prove (5.54), we first note that $r = 0$ makes no contribution because of the multiplication by r. For non-zero r, may cancel the r that is being averaged against the $r!$ in the denominator of (5.39) to get

$$\sum_{r=0}^{\infty} rP(r) = \sum_{r=1}^{\infty} \frac{\mu^r}{(r-1)!}e^{-\mu} \tag{5.57}$$

The second sum begins at $r = 1$ because of the remark just made. Now call $r - 1$ by the new name s, note that $r = 1, 2, 3, \ldots$ means $s = 0, 1, 2, \ldots$, and that μ^r is equal to $\mu \times \mu^s$ to get

$$\sum_{r=0}^{\infty} rP(r) = \mu \sum_{s=0}^{\infty} \frac{\mu^s}{s!}e^{-\mu} = \mu. \tag{5.58}$$

In the last step we have used the normalization condition (5.53) proved in (5.56).

Equation (5.55) can be proved in various ways. Here is a 'clever' proof which exploits a trick similar to the one just used. [Incidentally, there is no cook-book of recipes for such things. Just as we first discovered in Chapter 2, it is ultimately intuition and wasted hours that guide one.] Consider the average of $r(r - 1)$, namely $\sum_{r=0}^{\infty} r(r-1)P(r)$. Now, neither $r = 0$ nor $r = 1$ contribute to the sum, because $r(r - 1)$ is zero in these two cases. For r greater than or equal to 2, we can divide out this factor against the $r!$ in the denominator of (5.39). Then, giving $r - 2$ the new name t, and taking out a common factor of μ^2, we have

$$\sum_{r=0}^{\infty} r(r-1)P(r) = \mu^2 \sum_{t=0}^{\infty} \frac{\mu^t}{t!}e^{-\mu} = \mu^2 \tag{5.59}$$

Now, $r(r-1) = r^2 - r$, so that [compare (3.13)] (5.59) yields:

$$\sum_{r=0}^{\infty} r(r-1)P(r) = \mu^2 = \sum_{r=0}^{\infty} r^2 P(r) - \sum_{r=0}^{\infty} rP(r)$$

$$= \sum_{r=0}^{\infty} r^2 P(r) - \mu, \tag{5.60}$$

where we used (5.54) in the last step By this trickery, we have achieved the result

$$\sum_{r=0}^{\infty} r^2 P(r) = \mu^2 + \mu. \tag{5.61}$$

The variance given in (5.55) can also be expressed in the form given in the Appendix to Chapter 3.

$$\sigma^2 = \sum_{r=0}^{\infty} r^2 P(r) - \mu^2 = \mu^2 + \mu - \mu^2 = \mu. \tag{5.62}$$

Above, the first equality is the completely general expression (3.17), and the second equality comes from using (5.61). A cancellation has then occurred, leading to (5.55)!

6

Forces, motion, and energy

> ... the whole burden of philosophy seems to consist
> in this – from the phenomena of motions to investigate
> the forces of nature, and then from these forces to
> demonstrate the other phenomena ...
>
> Isaac Newton

Probability enters theoretical physics in two important ways: in the theory of heat, which is a manifestation of the irregular motions of the microscopic constituents of matter; and, in quantum mechanics, where it plays the bizarre but, as far as we know, fundamental role already briefly mentioned in the discussion of radioactive decay.

Before we can understand heat, we have to understand motion. What makes objects move, and how do they move? Isaac Newton, in the course of explaining the motion of planets and of things around us that we can see and feel with our unaided senses, answered these questions for such motions three centuries ago. The science he founded has come to be called classical or Newtonian mechanics, to distinguish it from quantum mechanics, the theory of motion in the atomic and sub-atomic world.

Classical mechanics is summarized in Newton's 'laws' of motion. These will here be illustrated by an example involving the gravitational attraction, described by Newton's 'law' of gravitation. In the brief description that follows, I shall attempt to show that these laws of nature: (i) unify apparently unrelated phenomena – like the motion of the moon and the fall of an apple; (ii) make simple but astonishing quantitative connections – between, for example, the rate of the moon's revolution around the earth and the numerical value of the acceleration in an apple's fall; and (iii), suggest new and useful concepts – here kinetic energy, potential energy, and, more generally, energy – which provide access to unexplored territory.

To say that an object moves is to say that its position in space changes with time. This can only be given meaning by referring it to some fixed point. Imagine a straight line directed *from* such a fixed point *to* the place where our object is. This directed line is called the position 'vector' of the object. The word vector is used to indicate something that has a magnitude and a direction. We shall encounter other vector quantities soon. Two vectors are said to be equal when they have the same magnitude *and* direction.

To give a quantitative description of movement, one requires units in which time and the magnitude of the position vector can be specified. No completely rational system of units exists: nature provides us with many more or less accurate clocks, from the rotating earth to the frequencies of atomic radiation, and many tape-measures, from the circumference of the earth to the wavelengths of those radiations. The system of units commonly in scientific use – and also in every day use in every country but the US – has a geocentric history. Time is measured in seconds (abbreviated as s), and one second was originally related, via a Babylonian process of subdivision still in use, to the average period of the earth's rotation, the day. Length is measured in meters (abbreviated as m), and one meter was originally defined as a ten millionth part of the distance from the North Pole to the equator measured along the great circle that passes through Paris. These picturesque definitions are no longer primary, but details of the current standards need not concern us: the internationally agreed upon meter and second are the historical units made more precise by tying them to measurements that can be done in a laboratory.†

The science of mechanics also requires a way of specifying the 'quantity of matter' in an object. The common scientific unit for this purpose is the kilogram (abbreviated as kg), originally the quantity of matter associated with – or, to use the technical term, the 'mass' of – one liter of water, and now an appropriate refinement of that definition. The system of units based on the meter, the kilogram, and the second is called the mks system; it is part of the SI, or *Système International*.

Newton's laws of motion

According to Newton, there are three laws of motion from which all observations about the mechanics of ordinary inanimate objects can be deduced. These laws connect concepts that can be quantified using the units just introduced.

Newton's first law states that motion at constant speed in a straight line, or as a special case no motion at all, occurs when there are either

† In metric units volume is often measured in liters. One liter is one thousandth of a cubic meter. Since a centimeter is one hundredth part of a meter, a liter is the volume of a cube 10 centimeters on a side.

no external influences, which he called forces, or a perfect balance between them. Thus, when an ice hockey puck skids along the ice, it does so at constant velocity (which means directed speed and will be defined more precisely soon) except to the extent that the small frictional effect of the surface slightly slows it down, and that collisions with other objects turn it right or left. When you or I sit in a chair we are at 'rest,' to use Newton's word for the condition of no motion, because there is a balance of forces – our weight vs. the support of the chair.

Newton's second law addresses the question of what happens when there is an imbalance of forces. For an object consisting of a fixed quantity of matter it states that 'acceleration,' or rate of change of velocity, is proportional to impressed force, the coefficient of proportionality being the mass. This sounds like a circular chain of definitions until one thinks about it. Imagine a device that produces a force of a certain amount, say a spring stretched a certain distance. This force applied to a given mass is found to produce a constant acceleration of a certain amount. The same force applied to twice the quantity of matter is found to produce half the acceleration. This is, thus, not definition but a summary of how nature works. To quantify the notion of force one must measure the other quantities in the units introduced above. Imagine one kilogram of matter being accelerated at one mks unit, a notion that will be made precise below. The force needed to do this is one unit of force. It is called, appropriately enough, one newton.

Newton's third law says that forces come in equal and opposite pairs: for every action there is an equal and opposite reaction, as he put it. It is instructive to think of examples of such action–reaction pairs. If I push against a wall, the wall exerts an equal and opposite force on me. What about a person sitting in a chair? Are the person's weight and the force from the chair on the person action and reaction? The answer is *no*. It is only when the person is at rest that this pair of forces are equal and opposite. If the chair is in an amusement park roller-coaster† where the occupant is being accelerated then there will be a net force on the person coming from the imbalance between weight and the force from the chair. A correct action–reaction pair is the force exerted by the person on the seat and the always equal and opposite force exerted by the seat on the person. To clarify this point consider a person sitting on a chair in a stationary room. What happens if the chair is suddenly pulled away? At first sight it

---

† See solved problems (1) and (3) at the end of this chapter.

seems that there is now only a single force, weight, which acts on the unfortunate person and causes him or her to be accelerated towards the ground. What the 'reaction' is to an object's weight will emerge in the section on Newton's Law of Gravitation. Weight is in fact the gravitational force of attraction between the object and the earth. It is exactly balanced by a reaction force attracting the earth toward the object. Why does the earth then not accelerate towards a falling object? In fact, it *does*. However, because of the huge difference in the masses of the earth and the person, the equal and opposite gravitational forces – a true action–reaction pair – lead to hugely unequal accelerations.

The only way to become truly comfortable with these ideas is to examine a large number of special cases and verify that they agree with experience. Unfortunately, there isn't time for that here. Instead, in the spirit of Chapter 1, we shall try to appreciate the power and generality of Newton's extraordinary synthesis by doing some interesting examples completely.

Two examples of acceleration

So far we have not properly defined either velocity or acceleration. Both acceleration and velocity are vectors, and are in this respect like position. Velocity is by definition the rate of change of position. We previously described velocity as 'directed speed.' We shall see shortly that this intuitive description corresponds exactly to the precise definition. Acceleration is defined as the rate of change of velocity. We see from its definition that acceleration is to velocity as velocity is to position. An object whose position is constant in time has no velocity; an object whose velocity is constant in time has no acceleration. These matters are best understood in special cases. We shall consider just two: motion along a circular path at a constant rate, and constant acceleration.

Two parameters are needed to specify motion on a circular path at a constant rate: the radius of the circle, call it r, and the period (or time taken to traverse the circumference), which we shall call τ, Greek lower case 'tau.' Since we are considering motion at constant speed, with no starts, stops, speed-ups, or slow-downs, this speed, v, is the circumference of the circle divided by the period. The circumference of a circle of radius r is $2\pi r$, which leads to the relation

$$v = \frac{2\pi r}{\tau}.$$

(6.1)

As calculated, v is the average speed associated with one circuit. The assumption of a constant rate of traversal implies that this is also the speed at every instant.

What is the velocity vector in this case? Let us first obtain the answer from the intuitive notion: directed speed. The magnitude of the velocity vector is then the speed just calculated. The direction of travel at any instant is tangent to the path. For a circle, the tangent is perpendicular to the radius. So that must be the direction of the velocity vector.

We can be more precise. Here, a drawing is of great help. Fig. 6.1 shows a part of the circular trajectory. We are contemplating the situation in which a particle steadily progresses around the circle as time passes. Imagine two nearby instants of time which we shall call t and $t + \Delta t$. The Δ (Greek capital 'delta') is a conventional mathematical usage for 'a little bit of.' Thus Δt means a small interval of time. At two nearby points on the circumference of the circle, the positions of the particle relative to the center are shown by arrows. These two positions have been labeled $\vec{r}(t)$ and $\vec{r}(t + \Delta t)$. The symbol $\vec{r}$ stands for the position vector. The notation $\vec{r}(t)$ is to allow for changes of position with time and stands for the position vector at

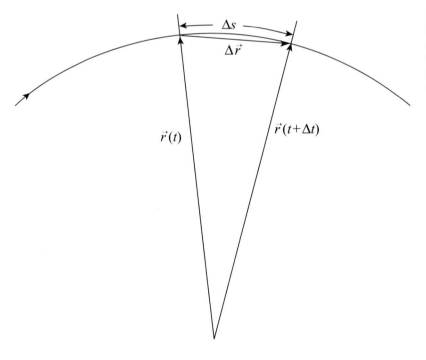

Fig. 6.1. Drawing illustrating circular motion. The particle is shown at two nearby instants of time.

time t. Correspondingly, $\vec{r}(t + \Delta t)$ stands for the position a small interval of time later. From the two arrows pointing to the positions of the moving particle at two nearby instants of time, one can define precisely what is meant by the velocity of the particle. The short arrow in the figure labeled $\vec{\Delta r}$ has been drawn in such a way that following $\vec{r}(t)$ from its tail at the center of the circle to its arrowhead and then following $\vec{\Delta r}$ from its tail to its head reaches the head of the arrow representing $\vec{r}(t + \Delta t)$. It is thus natural to think of $\vec{r}(t + \Delta t)$ as the sum of the two vectors $\vec{r}(t)$ and $\vec{\Delta r}$, and this 'triangular addition rule' is in fact the definition of vector summation. Correspondingly, $\vec{\Delta r}$ is defined to be the vector difference between $\vec{r}(t + \Delta t)$ and $\vec{r}(t)$. This is written

$$\vec{\Delta r} = \vec{r}(t + \Delta t) - \vec{r}(t). \tag{6.2}$$

Velocity was just defined as 'rate of change of position.' Since $\vec{\Delta r}$ is the small change of $\vec{r}$ in the small time Δt, the velocity vector is given by the expression

$$\vec{v} = \frac{\vec{\Delta r}}{\Delta t}. \tag{6.3}$$

Let us examine the magnitude and the direction of (6.3) for our example. The only quantity with a direction on the right hand side of (6.3) is $\vec{\Delta r}$ which must therefore specify the direction of the velocity vector $\vec{v}$. Looking back at Fig. 6.1, we notice that the direction of $\vec{\Delta r}$ depends on how close to each other are the points labeled with the times t and $t + \Delta t$. So the definition given by (6.3) depends on the size of Δt. We are here being forced to confront the difference between *average* and *instantaneous* rates.

A discussion of the difference between average and instantaneous speed may be a useful preliminary. When two runners finish in a dead heat, they have run their race at the same average speed, which does not imply that they ran neck and neck all they way, i.e. at the same speed as each other at every instant of the race. Presumably, also, neither ran at constant speed. Only in that case would the speed at every instant be equal to the average speed.

Equation (6.3) defines the average velocity in the interval Δt following t, in the same way as the distance run in a certain time defines the average speed of the runner. To associate an instantaneous velocity with the time t, one must imagine the limit of the right hand side of (6.3) as Δt is made smaller and smaller. Obviously, $\vec{\Delta r}$ then also becomes shorter and shorter. A glance at the figure shows that the limiting *direction* of $\vec{\Delta r}$, which through (6.3) is also the direction of $\vec{v}$ at the instant t, is indeed tangent to the circle.

What is the magnitude of the velocity obtained from (6.3)? If the length of $\vec{\Delta r}$ is much less than the radius of the circle, this length is closely equal to the length of the circular arc connecting the arrowheads of the vectors $\vec{r}(t)$ and $\vec{r}(t + \Delta t)$. This arc length is the fraction of the circumference traversed in a time Δt. By assumption the entire circumference $2\pi r$ is being traversed in the time τ. So the arc length, call it Δs, is given by

$$\Delta s = 2\pi r \frac{\Delta t}{\tau}. \tag{6.4}$$

If Δt is taken to be very small, Δs also becomes very small. At the same time, the arc length approaches the straight line segment, which is the magnitude of $\vec{\Delta r}$. Thus the ratio $\Delta s/\Delta t$ becomes equal to the magnitude of the velocity. Equation (6.4) then shows that the instantaneous velocity has the constant magnitude $(2\pi r/\tau)$ exactly as calculated in (6.1). The formal expression (6.3) thus gives the answers obtained intuitively at the start of this section.

The same argument shows quite generally that in the limit of arbitrarily small Δt the velocity vector defined by (6.3) has a direction tangent to the path of travel and a magnitude equal to the speed along the path at the instant of time t.

Now we move on to consider the acceleration. In Fig. 6.2(a) are shown not only the position vectors of our moving particle at two nearby times but also the two corresponding velocity vectors, represented by arrows with constant lengths and directions tangent to the circle. As the particle rotates around the center of the circle, we see that the velocity arrows also rotate and at the same rate.

The mathematical way of expressing the fact that the acceleration vector is the rate of change of velocity is a formula analogous to (6.3):

$$\vec{a} = \frac{\vec{\Delta v}}{\Delta t}. \tag{6.5}$$

Here $\vec{\Delta v}$ is the vector difference of the instantaneous velocities at times $t + \Delta t$ and t. In our special case the magnitude of the velocity does not change; it is always given by (6.1). Figure 6.2(a) shows that in a time interval Δt the direction of the velocity changes by exactly the same angle as the position vector. This is made clearer in Fig. 6.2(b) where the velocities $\vec{v}(t)$ and $\vec{v}(t + \Delta t)$ have been made to start at the same point without changing their directions. The vector difference $\vec{\Delta v}$ has also been explicitly drawn.

By working with the triangle of velocities in Fig. 6.2(b) exactly as we did with the triangle of positions in Fig. 6.1, it is possible to deduce the magnitude and direction of the acceleration of a particle moving along a circular path at a constant rate. But it is also possible to make an intuitive leap to the answer. Since the velocity vector is rotating in a circle at the same constant rate as the position vector, and since acceleration is to velocity as velocity is to position, an equation of the form (6.1) must give the magnitude of the acceleration, call it a, in terms of the magnitude of the velocity, v. Replacing v by a and r by v in (6.1), we obtain that the *magnitude* of the acceleration of a particle moving at a constant rate along a circular path is independent of time and given by

$$a = \frac{2\pi v}{\tau} = (\frac{2\pi}{\tau})^2 r,$$
(6.6)

where in the last form we have used (6.1) to eliminate v.

The direction of the acceleration vector can also be obtained by the analogy. Note from Fig. 6.2(a) that the direction of $\vec{v}(t)$ is a 90° right turn from the direction of $\vec{r}(t)$. A further 90° turn to the right must then connect $\vec{a}(t)$ with $\vec{v}(t)$.† This is the direction *opposite* to $\vec{r}(t)$. In short, a particle moving along a circular path at a constant rate

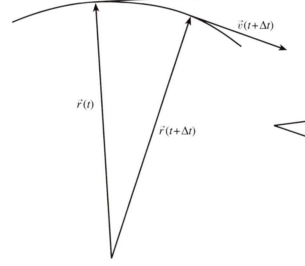

Fig. 6.2. Uniform circular motion. (a) The position and velocity vectors at two nearby instants of time. (b) Two nearby velocity vectors and their difference.

(a)   (b)

† Verified by examining Fig. 6.2(b) and noting the direction of $\Delta\vec{v}$, which is also through (6.5) the direction of $\vec{a}$, in the limit as Δt approaches zero.

is constantly being accelerated towards the center of the circle by the amount given by (6.6).

Note immediately that this is a situation to which Newton's first law does *not* apply. Although the speed of the particle is unchangingly constant, the direction of motion and therefore the velocity are constantly changing. The acceleration associated with the motion, which we have just calculated, implies via Newton's second law that a massive particle can only be made to move in a circle at constant speed by the application of a force pointing towards the center of the circle. The needed force is familiar to everyone who has whirled something around at the end of a string. (The assumption implicit in the previous statement is that other forces, such the air resistance of an object on a string – which would slow it down, – are small enough to be reasonably ignored.)

Now we are in a position to understand one unit of acceleration in an intuitive way. Suppose that an object is moving steadily around a circle of radius one meter at such a rate that it makes one circuit in 2π seconds, i.e. approximately 6.28 seconds. Then, using (6.6), we find that its acceleration towards the center is one mks unit – one meter/(second)^2. If the object has a mass of one kilogram, Newton's second law then informs us that this requires a steady force pulling it towards the center of one mks unit of force, or one newton.

The one other situation I would like to explain is easier to understand: constant acceleration in the direction of the velocity. This is, approximately, what happens when an object is dropped from a window. [Approximately because the gravitational force is not the only force acting on such a falling object; there is air resistance as well.] The force of gravity is proportional to the mass of the object; this force causes all bodies falling near the surface of the earth to accelerate downward at the same rate, which in mks units is approximately 9.81 m/s^2. An object released from rest thus picks up downward velocity as it accelerates.

The ubiquity of frictional forces opposing motions in our everyday world makes it easy to understand why astronomical observations played such an important role in the insights of Newton and his contemporaries. On the other hand, although the original impetus for the development of mechanics may have been the explanation of celestial motions, one of Newton's great triumphs was the unification of celestial and terrestrial dynamics. We are very close to being able to understand this synthesis. What follows is a small digression in that direction. Thereafter, I shall return to the main path of extracting those concepts from mechanics which allow one to understand the random motions which we experience as heat.

Newton's law of gravitation

Newton discovered that he could explain the elliptical trajectories of the planets around the sun, as well as other properties of these orbits systematized by Johannes Kepler [1571–1630], by assuming (i) that the sun attracted each planet with a force proportional to the inverse square of the distance between them, and (ii) that planets followed the dynamical laws we discussed in the last section. He then made the astonishingly bold assumption that every particle of matter in the universe attracts every other particle in the same way. This 'universal' law of gravitation can be written in the form

$$F \;=\; G\frac{m_1\, m_2}{r_{12}^2} \tag{6.7}$$

In this equation F is the magnitude of the gravitation force, m_1 and m_2 are the masses of the two particles, r_{12} is the distance between the two particles, and G is the so called gravitational constant. The direction of the force is along the line joining the two particles: particle 1 attracts particle 2 along this line, and vice versa.

Various motions are possible for two masses moving under the influence of their mutual gravitational attraction. Two cases that are well within our ability to understand completely are circular motion and a head-on collision. To keep everything as simple as possible, let us assume that one of the objects is very much more massive than the other. It is then a good first approximation to assume that the more massive object is not accelerating at all – because the accelerations produced by the equal and opposite forces are inversely proportional to the masses – and to take the center of this object as the origin of the position vector of the other mass. As an example of circular motion, we may think of the earth–moon system, because the moon is, in fact, in an almost circular orbit around the earth. For a head-on collision we may think of an apple falling to the earth. If these are two manifestations of the same basic laws of nature, then there must be connections between them. What are they?

The question is answered by comparing the law of gravitation (6.7) with Newton's second law of motion, and doing this for both the system of earth and moon and the system of earth and apple. Let M_e, M_m, and M_a denote the masses of the earth, moon, and apple, respectively; let r_m be the radius of the moon's orbit; and let r_e be the radius of the earth. Substituting these definitions into the general formula (6.7), one obtains the following expressions for the force, call it F_m, holding the moon in its orbit, and for the force, F_a causing the apple to fall earthward:

$$F_m = G\frac{M_e\,M_m}{r_m^2} \tag{6.8}$$

and

$$F_a = G\frac{M_e\,M_a}{r_e^2}. \tag{6.9}$$

Now the acceleration of the moon towards the earth is given by (6.6), with r replaced by r_m (the radius of the moon's orbit), and τ by the time, call it τ_m, of one revolution of the moon around the earth. The acceleration of the apple, also towards the earth, has the value g discussed in the last section. Newton's second law thus tells us that

$$F_m = M_m\,(\frac{2\pi}{\tau_m})^2 \cdot r_m \tag{6.10}$$

and

$$F_a = M_a g. \tag{6.11}$$

Equating the right-hand sides of (6.8) and (6.10) and cancelling out the common factor of M_m, doing the same for (6.9), (6.11), and the common factor M_a, dividing the two resulting equations by each other, and making a few rearrangements one obtains the quite remarkable relationship

$$g\ r_e^2 = (\frac{2\pi}{\tau_m})^2\ r_m^3. \tag{6.12}$$

Note that the quantities on the left pertain to the earth, and thus to an apple on the earth's surface, whereas those on the right have to do with the moon. Even three centuries after Newton's prediction of this connection, there is something breathtaking about its scope. But, before getting lost in the grandeur of it, we should ask some down to earth questions about the formula: How well does it work? How well, given that assumptions were made in obtaining it from the postulated general laws, should it work?

Here are the measured values, to three significant figures, of the symbols in (6.12): $g = 9.81$ meters/seconds^2, $r_e = 6380$ kilometers, $\tau_m = 27.3$ days, and $r_m = 384\,000$ kilometers. Using these numbers, the left side of (6.12) is, also to three figures, 3.99×10^{14} meters^3/seconds^2, whereas the right side is 4.01×10^{14} meters^3/seconds^2. (Note that, in order not to be comparing apples with oranges, a definite system of units is needed. Here, all lengths have been converted to meters, and all times to seconds.) Another way of expressing how well or badly (6.12) is obeyed by the numbers given above is to use the last three of them to *calculate g*. When this is done, g comes out to be 9.87 meters/seconds^2, which is to be compared with the measured value of 9.81 in the same units. These calculations show that the agreement is very good but not perfect.

Should it be better? There are at least three explanations for the discrepancy: the hypothesized general laws could be not quite correct; the calculation that led from these laws to the formula being tested could be faulty; the observational data could be inaccurate. In the present case, we all know the answer. If Newton's laws of gravitation and of motion were only correct to a few percent for the movement of ordinary objects on the surface of the earth, and for planetary orbits, contemporary engineering and space science, which uses these laws, would not work. Even though the explanation here is obviously the second one, the point still worth making is that in the progress of science there are often occasions where judgements of this kind have to be made, usually on the basis of imperfect information and in an intuitive way.

Without going into details, let me mention that the most significant correction to (6.12) comes from the finite mass of the earth. We assumed above that the moon is moving around the earth in a circular orbit. In fact, the earth's pull on the moon is balanced by a force from the moon on the earth. This can be shown to cause both the earth and the moon to orbit around a point not quite at the center of the earth, and to a calculable correction to our formula.

There are other corrections. The alert reader will have noticed a curious anomaly in the calculation we have done. We said that Newton assumed that every massive point particle attracts every other such particle according to the inverse square law. Although, on the scale of the other objects being considered here, an apple is small enough, the radius of the moon and, even more, that of the earth are by no means negligible. What we have assumed is that moon and earth attract other objects as though all their mass were concentrated at their centers. This is something that worried Newton quite a lot, and it took him some time to realize that if every particle of a *spherically symmetrical* object, i.e. one that is made up out of concentric spherical shells of constant mass density, is assumed to attract a particle external to it by the inverse square law, then, quite remarkably, the total force on the external particle comes out to be exactly what one would get if the total mass of the sphere were concentrated at its center. This fact makes the calculation we did largely correct, and in fact provides a striking internal test of the inverse square law, but does mean that a correction has to be worked out for the not quite spherical shape of the earth. When these and other corrections† are made, my astronomer friends tell me that everything works very well.

In the very interesting (6.12) which we have here been discussing the gravitational constant G played no role. It is worth pointing out that this constant can be measured in the laboratory‡ by measuring the force between massive objects. The constant is found to be 6.67×10^{-11} in mks units – force in newtons, distances in meters, and masses in kilograms. It is then fascinating that the equation for the acceleration due to gravity, g, obtained by eliminating F_a from (6.9) and (6.11),

$$g = G \frac{M_e}{r_e^2}, \tag{6.13}$$

allows one to calculate the mass of the earth from its radius, the acceleration due to gravity, and the gravitational constant! The answer is close to 6×10^{24} Kg.

Newton's mechanics not only explained the motion of ordinary objects on the earth, and of celestial bodies, but it left as a more general legacy the concept of Energy, which turns out to be central to a discussion of heat. To this we now turn.

Energy

When an apple is falling towards the earth, nothing about its state of motion appears, at first sight, to be remaining the same. Its velocity, after all, is increasing, and its height above the surface of the earth is decreasing as time passes. It is, however, possible to view the fall of the apple as a process in which something called energy is being changed from one form to another, while itself remaining constant. Energy is one of what are called 'constants of the motion' in Newton's mechanics. Abstract and perhaps artificial though this concept seems, energy has become a common thread that runs through all branches of physics. It is possible to view each great advance in our understanding of the physical world since Newton's time as the identification of a new form of energy. The statement that processes occurring in nature

---

† such as those for the not quite circular orbit and the more subtle (see Chapter 11) effect of the sun on the earth–moon system.

‡ The first measurement was done by Lord Cavendish (1731–1810) in 1798.

transmute energy from one form to another but never change it in total amount remains as true today as it was then.

The two forms of energy of a falling object are associated with its velocity and with its height above the ground. Can we quantify the thought that the decrease in height is in some way exactly compensated for by the increase in downward velocity? Consider Fig. 6.3. It shows the apple at two closely separated instants of time, which we label t and $t + \Delta t$. During the small interval of time Δt, the height of the apple decreases by a small amount. If h is the height at the time t, call $h + \Delta h$ the height at the slightly later time. Similarly, let v and $v + \Delta v$ be the downward velocities at the two times, the second being slightly greater than the first. The signs have been chosen so that the small changes Δt and Δv are positive, and the small change Δh is negative. How are these various quantities related to each other and to the acceleration due to gravity g?

From the definition of acceleration as the rate of change of velocity, and the property, discussed in the last section, that the downward acceleration due to the attraction of the earth near its surface is the known constant g, it follows that

$$\frac{\Delta v}{\Delta t} = g. \tag{6.14}$$

Multiply both sides of this equation by $v\Delta t$ to get

$$v\Delta v = gv\Delta t. \tag{6.15}$$

Fig. 6.3. An apple falling illustrated at two nearby instants of time

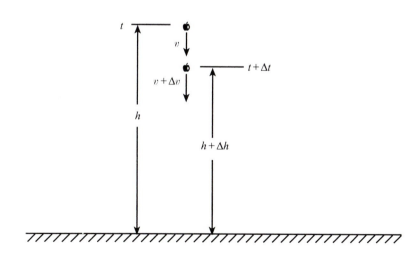

Now the downward velocity is equal to the rate of decrease of the height of the apple above the earth's surface, and thus

$$v = -\frac{\Delta h}{\Delta t}. \tag{6.16}$$

Using (6.16) in (6.15) and then moving both the terms in the latter equation to the left, we have

$$v\Delta v + g\Delta h = 0. \tag{6.17}$$

In order for the identification (6.14) to be correct it is necessary that Δv be very much smaller than v. [Otherwise $\Delta v/\Delta t$ would not be the instantaneous acceleration g.] Note that

$$(v + \Delta v)^2 - v^2 = 2v\Delta v + (\Delta v)^2$$
$$= 2v\Delta v \cdot [1 + \frac{\Delta v}{2v}]. \tag{6.18}$$

The left side of this equation is the small change in v^2 during the small interval time Δt. It should logically be written $\Delta(v^2)$. Since $\Delta v/v$ is very much smaller than 1, we thus have approximately

$$\Delta(v^2) = 2v\Delta v \tag{6.19}$$

and this equation becomes better and better as Δt becomes smaller and smaller. Thus (6.17) reads

$$\Delta[\frac{v^2}{2} + gh] = 0. \tag{6.20}$$

This last equation is telling us that for two extremely close intervals of time the quantity within the square brackets does not change. But, the same argument works for *any* two successive nearby instants of time. Since a finite interval of time is a succession of very small, or 'infinitesimal,' intervals we have the result that the sum of $v^2/2$ and gh remains unchanged as time progresses. This sum multiplied by the mass, call it M, of the falling object is called its energy, which we may call E.

Any falling body thus obeys the equation

$$\frac{1}{2}Mv^2 + Mgh = E \tag{6.21}$$

where E is a constant that depends on the height and velocity with which the fall started.

Suppose, for example, that an object falls from rest, i.e. with no velocity, from the height h_0. Then, from (6.21) it follows that the energy of the object is Mgh_0. The velocity at any other time is then related to the height at that time by

$$v^2 = 2g(h_0 - h) \tag{6.22}$$

where we have cancelled out a common factor of M and re-arranged terms. The solution of (6.22) is $v = \pm\sqrt{2g(h_0 - h)}$.† Here the positive sign refers to the increasing velocity as the body falls. Interestingly, there is also another solution, with the negative sign, which corresponds to an upwardly projected object slowing down and coming to rest at height h_0. Knowing that the energy is constant is thus a useful calculational tool.

Each of the two parts of the energy in (6.21) has its own name. The first term, $\frac{1}{2}Mv^2$, is called the *kinetic energy*; the second, Mgh, is called the *gravitational potential energy*. A falling object is converting potential energy into kinetic energy; an object projected upward indulges in the opposite transmutation until its kinetic energy – which, being proportional to the square of the velocity, cannot be negative – is used up. (The form of the gravitational potential energy that we have worked out applies only to objects near the surface of the earth; further out in space, it is given by a different expression.)

The calculation just done can be generalized to cover other than vertical motion. The result is that the expression (6.21) remains correct, but v^2 becomes the square of the magnitude of the velocity vector. In this more general case one needs a principle besides the constancy, or conservation, of energy to deduce the trajectory of a falling body. The needed principle follows from the fact that there is no force in the horizontal direction, and thus from Newton's laws no change in the horizontal component of the velocity. From these two 'conservation laws' it is possible to deduce that the trajectories are parabolas.

There are other forms of potential energy in nature. For example, consider a massive object at the end of a spring vibrating to and fro. Viewed in terms of the energy balance, this is a process in which energy is being passed back and forth between kinetic energy, for which the formula is the same as the one we just derived, and the potential energy associated with the stretch or compression of the spring.‡

---

† Equation (6.22) brings home the fact that massive objects subject only to gravitational forces all move in the same way, i.e. independently of their mass. This (experimentally confirmed) fact has come about because of the occurrence of the *same* M in two physically quite different contexts: Newton's Second Law, where it describes the inertia – or sluggish unwillingness to accelerate – of matter, and Newton's Law of Gravitation, where it describes the strength of the attraction to other massive bodies. The equality of 'inertial' and 'gravitational' mass is central to Einstein's General Theory of Relativity, in which it is also true that motion in a given gravitational environment is independent of mass.

‡ See solved problem (4).

What happens when our falling apple hits the earth? Obviously, *its* energy is no longer conserved. It now has the gravitational potential energy associated with being at rest on the earth's surface, but its kinetic energy has escaped somewhere. A careful measurement would show that the underside of the apple and the ground where it struck become very slightly hotter just after the impact. It is tempting to infer that the energy of bodily transport of the apple has been degraded into various small scale motions. It is a historical curiosity that this connection between dissipated mechanical motion and heat, obvious now to everyone who has noticed that a faulty bearing heats up, was precisely verified and accepted rather late in the development of the science of heat. Today, however, it is a truism that heat is chaotic motion on a microscopic scale. A profound understanding of heat can therefore be obtained by combining the notions of mechanical motion and randomness.

Even more than in previous chapters, the ideas presented here will be made clearer by working through, understanding, and thinking about the following solved problems. In each of them a massive object is subjected to forces. To apply Newton's Laws at any instant, one must set the mass times the acceleration equal in both magnitude and direction to the *net* force, obtained by vector addition (p. 85) of *all* the forces acting on the mass.† In the problems, the constancy of energy will emerge as an extremely useful calculational device.

Solved problems

(1) A person stands on a weighing machine in an elevator.
   (i) The elevator is descending at constant velocity. What does the machine register?
   (ii) The elevator is accelerating downward with an acceleration equal to one-tenth of the acceleration due to gravity, *g*. What does the machine register now?

   Solution: (i) A person moving at constant velocity has no acceleration, and therefore through Newton's first law has no unbalanced force. The forces on the person are (i) his or her weight acting downward, and (ii) an upward reaction force from the platform of the weighing machine. These must thus be equal and exactly cancel. From Newton's third law, the force on the platform has to be equal to the upward reaction force, which we just argued had to be equal to the person's weight. The machine will thus read true weight, exactly as it would if stationary.

---

† Perceptive readers will notice that all points in an extended body need not have the same acceleration. For example differences can arise from rotations of rigid objects. In these problems such rotations are negligible.

(ii) Since the downward acceleration is now $g\frac{1}{10}$ the net downward force through Newton's second law must be $Mg\frac{1}{10}$, where M is the person's mass. The net downward force is the weight minus the upward reaction force from the platform. Since Mg is the weight, the upward reaction must be $Mg\frac{9}{10}$. This through Newton's third law is the downward force on the platform. The machine will read $\frac{9}{10}$ths of true weight.

(2) A certain planet of mass M has two moons of masses m_1 and m_2 that move in *circular* orbits of radii r_1 and r_2, with periods τ_1 and τ_2 respectively. (The mass M is so much larger than either m_1 or m_2, and r_1 and r_2 are sufficiently different, that it is reasonable to neglect the gravitational force exerted by one moon on the other.)

  (i) What are the accelerations of the moons? In which directions do they point?

  (ii) What are the forces on the moons? In which directions do they point?

  (iii) Connect the quantities in (i) and (ii) using Newton's second law.

  (iv) Show that,

$$\frac{\tau_1^2}{r_1^3} = \frac{\tau_2^2}{r_2^3}$$

  [This is Kepler's third law of planetary motion as applied to circular orbits.]

  (v) Let the planet be earth and one of the moons be our moon, which is 384 400km away, and has a period of 27.3 days. Let the second 'moon' be a satellite in a geosynchronous orbit – one that stays at a fixed point above the earth as it rotates. How far is this satellite above the earth's surface? (Note that to be accurate, one must know that the earth's radius is 6400km, and suppose that it behaves like a point mass concentrated at its center.)

Solution: (i,ii) We know that for a circular orbit at constant angular velocity the acceleration is towards the center of the circle. The gravitational pull of the planet is also towards the center. Thus Newton's second Law is immediately satisfied as far as the directions of the forces and accelerations are concerned. [The most general orbit is an ellipse, with the center of attraction at one of its foci. In the general case, the motion is *not* at constant angular velocity, but just what is needed to make the acceleration point towards the source of the attraction.]

(iii,iv) Newton's second Law leads to Eqs. (6.8) and (6.10), with appropriate notational modifications:

$$F_{m_1} = G\frac{Mm_1}{r_1^2} = m_1\left(\frac{2\pi}{\tau_1}\right)^2 r_1, \tag{6.23}$$

with an identical equation for the second moon, in which case the subscript 1 is replaced by 2. One then sees that

$$GM = \left(\frac{2\pi}{\tau_1}\right)^2 r_1^3 = \left(\frac{2\pi}{\tau_2}\right)^2 r_2^3, \tag{6.24}$$

from which the required identity follows.

(v) A geocentric satellite has a period of one day. The radius of its orbit is easily calculable from the proved identity, using the information given about the moon, with the following result:

$$r_2 = \left[\left(\frac{\tau_2}{\tau_1} \right)^2 r_1^3 \right]^{\frac{1}{3}}$$

$$\approx 42\,400 \text{km}$$

(6.25)

To get the height above the earth's surface, we must subtract the earth's radius, leading to a height approximately equal to 36 000km

(3) An amusement park roller-coaster – see Fig. 6.4 – is designed to 'loop the loop' in such a way that at the instant when the passengers are upside down they experience the same force from the seats that they did when at rest. The aim of this problem is to calculate the height from which the train must start to roll in order to achieve this design goal. We shall assume that the wheels are extremely light, and that friction from all sources – air resistance, wheel bearings, etc – is negligible. These assumptions mean that each car of the train behaves like a point particle following the track, experiencing a vertical gravitational force, and a reaction force perpendicular to the track which has no effect on motion along the track.

  (i) Give an argument to show that the design requires that the total downward force on an object in the moving train at the top of the loop is *twice* its weight, Mg, where M is its mass.

 (ii) Then, use Newton's second law to calculate the square of the train's velocity at the top of the loop in terms of R, the radius of the loop, and g, the acceleration due to gravity.

(iii) Use conservation of energy to calculate the velocity at the bottom of the loop, and, hence, the required height H.

(iv) Using the velocity at the bottom of the loop, evaluated in (iii), calculate the force on a passenger due to the seat at the bottom of the loop.

 (v) If we want to design a roller-coaster that gives the thrill of 'weight-lessness' at the top of the loop, (no force on the passengers due to the seat), what must H be now?

Solution: It should be pointed out right away that this is *not* a situation in which the angular velocity around the loop is constant, because the gravitational force provides a constant downward acceleration. However, at the very top and very bottom of the loop, all the forces are vertical, and, therefore, so are the accelerations. (i) When the roller-coaster is at rest on the ground, the upward force from the seat exactly compensates for, and is thus equal and opposite to, the weight of the passenger. [See problem (1).] By design, the downward force from the seat at the top of the loop, is then also to be equal to the passenger's weight, with the result that the total force on the passenger will be this reaction force plus the weight, both now acting in the same downward direction. At the top of the loop the downward force is thus twice the passenger's weight, i.e. $2Mg$.

(ii) For motion around a circle at constant angular velocity, we have, by eliminating the period τ in the equations connecting acceleration with speed and speed with radius, (6.1) and (6.6), $a = v^2/r$ where a is the magnitude of the acceleration, v is the speed, and r is the radius. This expression applies at the top and the bottom, where the acceleration is perpendicular to the velocity. Newton's second Law at the top of the loop now gives $2Mg = Mv_t^2/R$ for every mass M in the roller-coaster, where the subscript t refers to 'top.' The desired relation, in which M does not appear is: $v_t^2 = 2gR$. (iii) At the bottom of the loop, the train will have converted the gravitational potential energy corresponding to the height $2R$ into additional kinetic energy along the track. Thus, v_b, the velocity there, is determined by conservation of energy to be given by

$$\tfrac{1}{2}Mv_b^2 = \tfrac{1}{2}Mv_t^2 + Mg \times 2R \tag{6.26}$$

which gives $v_b^2 = 6gR$. Since we are told that energy dissipation by friction or air resistance is negligible, v_b must also have been the speed after the initial descent. Conservation of energy thus gives

$$MgH = \tfrac{1}{2}Mv_b^2, \tag{6.27}$$

from which we conclude that $H = 3R$ is the required height. (iv) The upward acceleration at the lowest point is $v_b^2/R = 6g$. Here the reaction force from the seat is in the opposite direction from the weight. Thus, reaction force minus weight must be equal to the mass times the just calculated acceleration, leading to the conclusion that the reaction force is 7 times weight. [Fairly unpleasant, I would think.]

(v) For there to be no reaction force from the seat at the top, one would need $v_t^2/R = g$. Carrying through the argument given for part (iii) shows that the required height is now $\tfrac{5}{2}R$.

(4) We know that a particle moving at constant speed in a circle has an acceleration that is in the opposite direction to its position vector, and thus that a massive particle moving in this manner requires a force pointing toward the center.

In the following problem we are going to deduce something new from this knowledge. We shall exploit the fact that if two vectors $\vec{A}$ and $\vec{B}$ are

Fig. 6.4. Schematic illustration of roller-coaster.

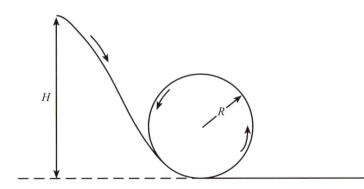

H

R

equal, their projections A_{pr} and B_{pr} in any direction must also be equal. This is illustrated in the Fig. 6.5(a)

Uniform circular motion is illustrated on the right of Fig. 6.5(b). The vertical projection of this motion is indicated in the left part. Notice that the projected motion corresponds to an up and down 'oscillation' between the points N and P. Such motion along a straight line is called 'simple harmonic motion.' We are going to see that such straight line oscillatory motion results from the application of a force proportional to and opposing the displacement from the point O. An example of such a system is a mass on a spring – the spring exerts a restoring force proportional to the amount the spring is extended or compressed from its length when at rest. In the left part of the figure such a spring is sketched. Note that the circular 'planetary' motion and the linear oscillation are two quite different things; each obeys Newton's equations in its own way. Projection is a device being used here to show that having understood the first motion allows one to deduce facts about the second.

(i) For the particle moving at constant speed in a circular orbit of radius R, show that there is a linear relation between the force F_{in} which acts in the inward direction and R *if* the period τ is required to be independent of R.

(ii) Since the vertical projection of the motion requires the application of the vertical projection of the force considered in part (i), show that the restoring force exerted by the spring must be given by $F = -kx$, where x is the displacement from the rest position O, and k is a coefficient called the spring constant. This implies that a mass on a such a spring undergoes simple harmonic motion. Arguing that F

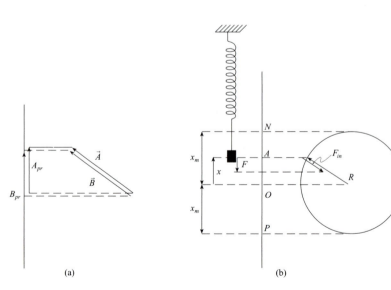

(a)   (b)

Fig. 6.5. (a) Illustration of vector projection; (b) Simple harmonic motion as a projection of circular motion.

and x must be related in the same way as F_{in} and R in (i), find the period of the oscillations in terms of k and the mass M.

(iii) Using the known relation between speed v and radius R in the circular motion case, relate the speed of the mass on the spring at point O – at which point the circular motion is vertical – to the maximum displacement of the spring, x_m.

(iv) Use the result in (iii) to show that the kinetic energy of the particle at O is $kx_m^2/2$. This is also the *total energy*, if we take the potential energy of the spring to be zero at its rest position O. At point N, where the speed up or down is zero, all energy is potential, from which it follows that the potential energy here is $kx_m^2/2$. Thus the potential energy due to an extension x is $kx^2/2$, and conservation of energy for the system gives,

$$\frac{1}{2}Mv^2 + \frac{1}{2}kx^2 = \text{constant} = \frac{1}{2}kx_m^2, \tag{6.28}$$

where the constant is the total energy. The condition (6.28) allows us to find the speed at any point if we know the maximum displacement.

Solution: (i) At constant angular velocity, we have calculated the inward acceleration [Equation (6.6)] to be $(2\pi/\tau)^2 R$. To keep a particle moving in this way an inward force F_{in} equal to the mass of the particle times this acceleration is required. If τ, the period of the motion, is to remain constant, F_{in} must be proportional to R.

(ii) Since the circular motion described above is consistent with Newton's Laws, its vertical projection must also so be. This proves that a restoring force linear in the vertical dislacement x gives rise to a to-and-fro motion identical to the vertical projection of uniform circular motion. Comparing the assumed $F = -kx$ with the vertical projection of the F_{in} calculated in (i), we conclude that $k = (2\pi/\tau)^2 M$. In short, a mass M attached to a spring with spring constant k, performs simple harmonic motion with a period $\tau = 2\pi\sqrt{M/k}$, regardless of the maximum extent ('amplitude') of the vibration.

(iii) We know that the speed in the circular orbit is always $(2\pi/\tau)R$. The maximum vertical projection occurs at $x = 0$, where it is equal to the speed in the circular orbit. Comparing the two problems, we see that the maximum velocity for the spring driven motion is $v_m = (2\pi x_m)/\tau$, where τ was calculated in part (ii).

(iv) Squaring the last equation and substituting for τ, we obtain: $\frac{1}{2}Mv_m^2 = \frac{1}{2}kx_m^2$. Since the left side is the maximum kinetic energy, the right hand side, corresponding to the situation where the velocity, and thus the kinetic energy, is zero must be the potential energy corresponding to a stretch x_m. Conservation of energy thus has the given form, allowing one to calculate the speed at any point if the maximum displacement is known.

7

Atoms, molecules, and molecular motion

> If, in some cataclysm, all of scientific knowledge were to be
> destroyed, and only one sentence passed on to the next
> generation of creatures, what statement would contain the
> most information in the fewest words? I believe it is the
> *atomic hypothesis* (or the atomic *fact*, or whatever you
> wish to call it)...
>
> Richard P. Feynman

That a gas, air for example, is made up of a myriad tiny bodies, now called molecules, moving about randomly and occasionally colliding with each other and with the walls of a containing vessel, is today a commonplace fact that can be verified in many ways. Interestingly, this molecular view was widely though not universally accepted long before there were experimental methods for directly confirming it. The credit for the insight has to go to Chemistry, because a careful study of chemical reactions revealed regularities that could most easily be understood on the basis of the molecular hypothesis. These regularities were known to chemists by the end of the eighteenth century. By this time, the notion of distinct chemical species, or 'elements,' was well established, these being substances, like oxygen and sulfur, that resisted further chemical breakdown. It was discovered that when elements combine to make chemical compounds they do so in definite weight ratios. It was also found that when two elements combine to make more than one compound the weights of one of the elements, when referred to a definite weight of the second, stand one to another in the ratio of small integers – which sentence is perhaps too Byzantine to be made sense of without a definite example.

Consider the oxides of sulfur. There are two of them, both gases at ordinary temperatures and pressures, and neither very pleasant to breathe; both implicated in acid rain, where they do not belong, and important in chemical engineering, where they do. We can learn something from them. They contain, respectively, 50.05 % and 40.04 % by weight of sulfur: the ratios of sulfur to oxygen in these compounds are $f_1 = (50.05/49.95)$ and $f_2 = (40.04/59.96)$. Amounts of the two compounds containing a given weight of oxygen thus contain sulfur in the ratio f_1/f_2. Do you agree? Work out this ratio, and you will find to two decimal places, which is the accuracy of the data, the number 1.50. In short, when sulfur combines with oxygen it can do so in two ways: if two parts of sulfur combine with a certain amount of oxygen to make one of the oxides, three parts of sulfur combine with the same amount of oxygen to make the second one. The emergence of these small integers from genuinely uninspiring raw data is a strong indication that some simple principle is at work.†

Regularities of this kind led the English chemist John Dalton (1776–1856) to propose the *atomic* and *molecular* hypotheses in the early 1800s. The suggestion was that elements consist of extremely small particles called atoms, which act as units in chemical and physical processes, and that chemical substances also have smallest units, now called molecules, which are aggregates of atoms, usually small numbers of atoms. A chemical formula tells how many atoms of different elements are combined in a single molecule of a compound. We know now that the oxides of sulfur are sulfur dioxide, written SO_2, and sulfur trioxide, written SO_3. [The subscripts denote the number of atoms in the molecule – 'one' being understood.] A little thought will convince you that, although these chemical formulas are compatible with the weight fractions given at the end of the last paragraph, some more information is needed before one can exclude such possibilities as S_3O for the sulfur-rich oxide, and S_2O for the sulfur-poor one. The missing information is the weight of the sulfur atom relative to that of the oxygen atom. If we knew that sulfur atoms are a little over twice as heavy as oxygen atoms, which is in fact true, then, on the basis of the atomic and molecular hypotheses, the oxide containing 50.05 % sulfur would have to have the formula SO_2, or, possibly S_2O_4. Chemists early learnt that the relative weights of atoms are contained in the concept of *combining weight*. It was found that a weight unit could be assigned to each element in such a way that when elements combine to form pure chemical substances, only small whole number multiples of these weights are involved. The standard for these combining weights was originally taken to be 16 grams of Oxygen.‡ On this scale the combining weight of sulfur was known to be 32.06 by the middle of the nineteenth century, thus fixing the relative number of atoms of

---

† No straightforward manipulation of equally uninspiring, but equally important, raw data in economics – the number of car-loadings of cheese exported from Denmark in different months, for example – yields ratios of small integers, which indicates that any simple underlying mathematical principles in the 'dismal science' are well hidden. [See Chapter 11 for a modern perspective on how such things can happen.]

‡ The modern scale of atomic weights is based on assigning the numerical value 12 to the form of the Carbon atom whose central core, or 'nucleus,' contains 6 'protons' and 6 'neutrons.'

sulfur and oxygen in the oxides of sulfur, as just discussed. [For future reference, it is worth noting that on the oxygen scale of combining weights, the number associated with hydrogen is 1.008.]

According to the atomic hypothesis, then, one combining weight of every element contains the same number of atoms. How many? The question occurred more than once to thoughtful scientists during the latter half of the nineteenth century. Several ingenious ways were suggested of determining this number, called Avogadro's number – after Amadeo Avogadro (1776–1818), who discussed in 1811 how it might enter measurements on gases that we shall discuss soon. Here is a thought experiment which allows one in principle to obtain this number. It is based on some real experiments done around 1890 in Cambridge by the famous English physicist Lord Rayleigh (1842 –1919), and by Miss Agnes Pockels, who wrote a letter from Brunswick (in Germany) to Rayleigh, which he forwarded to the editor of the journal *Nature*,† and referred to several times in his later work. Suppose one were to spread a carefully weighed small amount (a few drops) of a suitable oil of known molecular formula on the surface of water. If the material is properly chosen it will spread into a very thin layer, and it is known that there is a maximum size to the oil patch one obtains with a given amount of oil. Thinking of the molecules as little objects that both like to stick to each other and are also attracted to the water surface, the natural hypothesis is that the maximum obtainable area of continuous coverage corresponds to one molecular layer. Suppose that the molecules have linear dimensions a, and that they arrange themselves on the surface touching each other in a square pattern. These assumptions are not justified, and in fact we now know that they are not quite right, but if the overall picture of a molecular 'monolayer' is correct the deficiency can only result in an error of no more than a factor of about ten in what we may expect to be a very, very large number. Pushing on boldly, we would estimate that N molecules would occupy an area A given by

$$A = N\ a^2. \tag{7.1}$$

Suppose also that before spreading the oil on the water we measured its volume V. Assuming, for lack of anything better, that the molecules are arranged in a cubical pattern in the liquid, one has the equation

$$V = N\ a^3. \tag{7.2}$$

---

† 'I shall be obliged if you can find space for the accompanying translation of an interesting letter which I have received from a German lady, who with very homely appliances has arrived at valuable results …'

Now, by assumption, we know the molecular formula of the oil from clever chemical measurements, so we can easily work out the sum of the combining weights of the atomic constituents. The numerical value of this sum is called the *molecular weight* of the chemical compound. A quantity of oil whose weight in grams is one molecular weight must contain Avogadro's number of molecules. [Think about that for a second.] Let the (carefully measured) weight of the small amount of oil be a fraction f of the known molecular weight of the oil. Then, according to the molecular hypothesis the number of molecules, N, in our sample of oil is the same fraction f of Avogadro's number, which we shall call N_{Av}, i.e.,

$$N = N_{Av}f. \qquad (7.3)$$

Since A, V, and f have been measured in our thought experiment, (7.1-3) determine the unknowns N_{Av} and a as follows. Dividing (7.2) by (7.1) yields an expression for a, and dividing the cube (third power) of (7.1) by the square of (7.2) gives an expression for N, and thus via (7.3) for N_{Av}. The results are

$$a = \frac{V}{A}, \qquad N_{Av} = \frac{A^3}{V^2f}. \qquad (7.4)$$

So, not only does one get information about the size of the oil molecule from this experiment, one also gets an estimate for Avogadro's number!

Other more accurate, more direct, and more modern ways of measuring Avogadro's number exist. There is, for example, a method based on measuring the quantity of charge that is transported from one plate of an electrical cell to the other when a given amount of matter is so transferred, assuming we know the basic unit of charge, as we do. Other methods involve the use of probes, such as X-ray and neutron beams, which are waves with a crest to crest distance on the same scale as the size of atoms. I have described the Pockels–Rayleigh method because of its extreme conceptual simplicity. As an episode in the history of science, the story of the molecular hypothesis is fascinating in the lapse of about 85 years between Dalton's hypotheses and quite indirect measurements of the sort just described. Nonetheless, the circumstantial evidence for the picture, obtained mainly, as I have already said, from Chemistry, was so strong that it was an essential part of the way a majority of scientists conceived of matter in the latter half of the nineteenth century – even though they did not know that Avogadro's is the almost unimaginably large number 6.022×10^{23}, and that the characteristic size of atoms is the almost unimaginably small distance 10^{-8} centimeters. One hundred million atoms laid end to end would stretch not from here to China but only across a fingertip.

Molecules in motion: gases

Ice, water, and steam are the solid, liquid, and gaseous states of a substance familiar to us all. The gas is the most dispersed, least confined state. Does the molecular hypothesis lead to a mental picture of steam, and does this picture allow the calculation of anything? The answer to both questions is yes.

Consider a closed container empty except for a little water. If we heat the container, there comes a point at which the water boils away. With further heating, the pressure inside the container increases. The natural explanation is that the atoms that were contained in the liquid water are now filling the container. Are the atoms in the gas combined in the form of molecules? The chemical formula for water, obtained from the known fact that 2 grams of hydrogen combine with 16 grams of oxygen to give 18 grams of pure water, and from the combining weights of oxygen and hydrogen discussed above, is H_2O, or some multiple thereof, like H_4O_2. Which is it? Astonishingly, by taking the molecular hypothesis seriously it is possible to *deduce* from observations in the large that 18 grams of steam contain Avogadro's number of little objects – and not a half or a third as many. From this we conclude that steam consists of molecules of H_2O.

We may reason as follows. *Assume* that a gas is made up of molecules whose average kinetic energy is much larger than their average potential energy, the latter being due to the gravitational force and the weak forces of attraction between molecules which underly the condensation of gases into liquids.† These molecules are then flying about in the container. *Assume* that the motion of the molecules is governed by Newton's laws. Between collisions, with each other and with the walls of the container, the molecules are moving in straight lines, gravity, by hypothesis, being a very small correction. When a molecule bounces off the walls of the container, its velocity changes, i.e., it is accelerated. No acceleration without a force, says Newton. So the containing wall exerts a force on the molecule as it changes direction, and correspondingly, the molecule exerts a force on the wall. Gas pressure, then, according to the molecular hypothesis, is due to the continual bombardment of the walls by molecules. Why,

† See (6.21) for the definition of kinetic and gravitational potential energy.

then, does the pressure of a gas on containing walls not fluctuate? It *does*, but because there are so many molecules the law of large numbers comes in, and the fluctuating pressure is dominated by the mean.

Let us try to calculate the average pressure due to molecular collisions. First consider just one molecule. In general, it will approach the wall at some angle, so that its velocity has projections both perpendicular and parallel to the wall. A collision with a perfectly smooth and rigid wall will result in the reversal of the perpendicular component of the velocity, with no change in the kinetic energy – and thus the speed – of the molecule. As Fig. 7.1. shows, the component of velocity parallel to the wall is then unchanged. While the collision is taking place, the wall will experience the force discussed above.

Because of the enormous number of molecules within our container, it will be subjected to many such collisions. We can use Newton's second law to calculate the *average* force per unit area – which in technical usage is called the *pressure* – on the walls. Consider all the molecules with velocity perpendicular to the wall $v_\perp$. Let $\tau(v_\perp)$ be the average time between collisions of such molecules with a segment of the wall. Take M to be the identical masses of the molecules in a pure gas, to which we restrict ourselves for simplicity. Then the average force on the wall-segment, due to molecules with the perpendicular

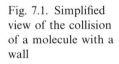

Fig. 7.1. Simplified view of the collision of a molecule with a wall

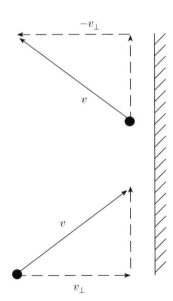

velocity being considered, is

$$F(v_\perp) = M \frac{2v_\perp}{\tau(v_\perp)}.$$ (7.5)

The idea being used here is that the average force is the mass times the average acceleration, the later being calculated using the expression (6.5). [Note that the change in velocity here is $2v_\perp$ and perpendicular to the wall (the 2 being due to the reversal in direction), which is divided by the average time interval between collisions, $\tau(v_\perp)$, to obtain the average acceleration.]

The time between collisions is the reciprocal of the average number of collisions per unit of time, and this number can be calculated as follows. [See Fig. 7.2.] Imagine a plane surface of area A parallel to the wall, moving towards the wall with velocity $v_\perp$, and just so far away from it now that it will reach the wall in a time Δt. In this interval of time, the imagined area sweeps out a cylinder of base A and length $v_\perp \Delta t$, i.e. of volume $Av_\perp \Delta t$. Now, every molecule with perpendicular velocity $v_\perp$ moves at the same rate as our imaginary surface.

It follows that every molecule of this kind in the cylinder just described collides with the wall in the time Δt. Let $n(v_\perp)$ be the average number of molecules per unit volume moving towards the wall with the velocity $v_\perp$. The reasoning just given says that the number of such molecules that collide with the wall in time Δt is $n(v_\perp)Av_\perp \Delta t$. This product must thus equal one when Δt is equal to $\tau(v_\perp)$, the average time in which one such collision takes place. So we have the connection

$$\frac{1}{\tau(v_\perp)} = Av_\perp\, n(v_\perp).$$ (7.6)

Substituting this into the previous formula, we get for the average force on the area A:

$$F(v_\perp) = 2M\, v_\perp^2\, n(v_\perp)\, A.$$ (7.7)

Finally, to find the total force, and from it the total pressure, we have to add the contributions of all the molecules moving toward the wall and divide by A. The answer for the average pressure p is

$$p = n\, M\, \langle v_\perp^2 \rangle$$ (7.8)

where n is the average number of molecules, regardless of velocity, per unit volume, and the angular bracket symbols $\langle\, \cdot\, \rangle$ enclose something which is to be averaged over all the molecules in the system. The only slightly mysterious thing about the transition from (7.7) to (7.8) is the disappearance of a factor of 2. This has occurred because n is the total density of molecules, whereas only those traveling to the right, i.e. on the average half the total, contribute to the pressure on the right wall. In short, the factor $n\,M$ is in more detail $(n/2) \times 2M$, which redisplays the $2M$ in (7.7). If we wish we may write the average density of molecules as the total number of molecules, call it N, contained in the volume of the container, call it V. Recalling the formula derived at the end of the last chapter for the kinetic energy of a particle, namely, $\frac{1}{2}Mv^2$, we thus have the result: the pressure of a gas in a container multiplied by the volume of the container is equal to the number of molecules in the container times twice the average kinetic energy per molecule associated with motion towards or away from the wall. In fact, since all three directions of space within the container are perfectly equivalent for the molecules, gravity being unimportant for them because of their assumed large kinetic energy, twice the average kinetic

Fig. 7.2. Cylindrical volume enclosing molecules that will strike a wall in the time interval Δt

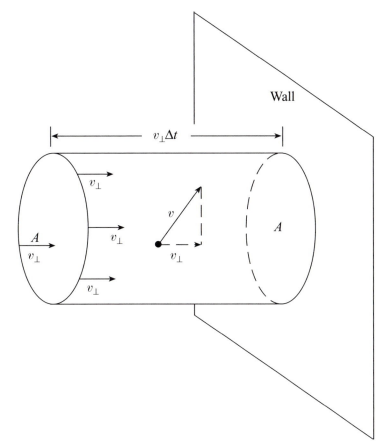

energy associated with motion towards or away from the wall is equal
to two-thirds of the average kinetic energy per molecule, without restrictions as to direction of motion. The formula just stated in words is

$$p \, V \;=\; N \, \frac{2}{3} \, \langle \tfrac{1}{2} M v^2 \rangle. \tag{7.9}$$

If we could find some way of measuring the average kinetic energy
per molecule of a gas, then, since the pressure p and the volume V
are easy enough to measure, we have, if the formula is correct, a way
of calculating the number of molecules in the container!

How 'correct' is the formula? To put the question another way, how
well, and under what circumstances, are the approximations made in
obtaining it justified? The key assumption has been that in calculating the pressure on a wall the kinetic energy of molecular motion
is more important than all the other energies in the problem. We
have neglected the forces that bind molecules together to make liquids and solids, and ignored those that bind atoms together to make
molecules. The first of these approximations must become better as
the gas becomes more *dilute*, because the molecules are then less likely
to encounter each other as they randomly move about.† As for the
forces that keep one molecule intact, we may argue that on the average they do not affect collisions with container walls. They are also
'frozen out' at ordinary temperatures by a quantum mechanical effect
to be discussed on p. 220.

At least one other approximation has been made. We assumed
that, on the average, the gas molecules uniformly fill the container
space, and are equally likely to be moving in every direction. This
will certainly not always be the case. For example, if we filled the
container with a powerful jet from a tank of compressed gas, the
molecules would initially be mainly moving in a definite direction. We
have assumed that any such motion is transient and has been allowed
to die out before the pressure has been observed, so that the gas has
reached a state which, as far as averages over many molecules are
concerned, does not change with time. In such a condition, the gas is
said to be in *equilibrium*. The assumption of equilibrium also justifies
the rather innocent way in which we treated the walls. Since a wall is
also made of atoms, it must be quite wrong to assume that it appears
smooth to a colliding molecule. However, the only effect of this crude
approximation was to allow us to ignore velocity changes parallel to

---

† In solved problem 3 (i) it is shown that under typical conditions in a dilute gas the average volume
per molecule is very much greater than the size of a molecule, thus justifying the assumption that
collisions between molecules are rare.

walls, and exchange of energy with the walls. Such effects will occur in single collisions, but in equilibrium they must average out to zero.

Equation (7.9) should thus apply to dilute gases in equilibrium,† and experiments on dilute gases should allow us to penetrate its secrets.

Molecular motion and gas temperature

How hot is something? Take its temperature. How? Use a thermometer. Temperature is hotness quantified. Matter in equilibrium, hot tea in a thermos flask, for example, has a definite temperature. Pour cool milk into the hot tea, and after some time a new temperature, higher than it was for the milk and lower than it was for the tea, is reached. Temperature is the concept missing in our discussion of molecules in motion, and we shall see that the molecular hypothesis gives vivid insight into this abstract idea.

The household mercury-in-glass clinical thermometer – which exploits the fact that the increase in volume of a fixed amount of mercury is proportional to the increase of temperature – is perhaps the most common device for measuring temperature. In this instrument temperature is read by comparing the length of the mercury column to a scale on the surface of the glass. In the US the scale is typically marked in ° F, for Fahrenheit; in the rest of the world it is marked in ° C, for Celsius. Neither of these scales is based on fundamental physical principles, but the latter has the advantage of attaching round numbers to two easily achieved temperatures: 100 to the boiling point of water at atmospheric pressure (the 'steam point,') and 0 to the freezing point of water under the same conditions (the 'ice point.')

A more fundamental temperature scale is obtained by using a remarkable property of *all* <u>sufficiently dilute</u> gases. Consider a quantity of gas confined to a given volume. The pressure exerted by the gas on the walls of its container increases as the temperature is increased. One finds that <u>the ratio of the pressures corresponding to two temperatures</u> – take them to be the steam point and the ice point for definiteness – <u>is independent of the particular gas</u>.

This experimental fact can be demonstrated in a simple device called a constant volume gas thermometer. It consists of a glass bulb in which gas can be trapped, and a system of glass and flexible tubing containing mercury which can be adjusted to force the gas to occupy a marked volume in the bulb. [See Fig. 7.3.] The pressure of the gas in

---

† Dilute mixtures of pure gases are also covered, because the calculation could be done separately for each molecular species in the mixture.

the bulb is related to the height of the mercury column as follows. The weight – the Mg force that caused the apple to fall in the last chapter – of the mercury in the middle column above its level in the left is ρhAg, where ρ (Greek 'rho') is the mass per unit volume of mercury, h is the height difference, A is the cross-sectional area of the tube, and g is the acceleration due to gravity. This weight exactly compensates for the difference between the forces due to the pressure of the gas, p, and the atmospheric pressure, p_{atm}, acting on the open end of the tube. Since these forces are the respective pressures multiplied by the area we have called A, and A can now be cancelled out everywhere, we have the result

$$p = p_{atm} + \rho g h, \tag{7.10}$$

so that the gas pressure is easily calculated from the measured height difference h of the mercury columns.

In the experiment we are contemplating the gas-containing bulb is

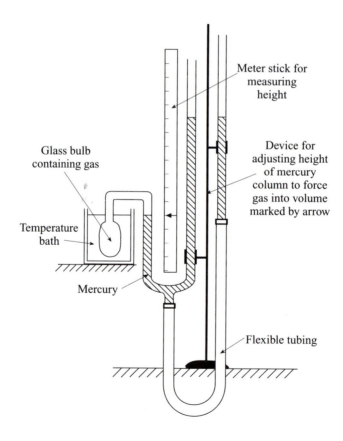

Fig. 7.3. A constant volume gas thermometer.

Meter stick for measuring height

Glass bulb containing gas

Device for adjusting height of mercury column to force gas into volume marked by arrow

Temperature bath

Mercury

Flexible tubing

successively put into baths at the steam and ice points and the pressure of the marked volume is recorded at both temperatures; some gas is then removed, the two pressures measured again, and the procedure is repeated. Typical data for an experiment of this kind performed on two different gases are shown in Fig. 7.4, which illustrates the empirical fact underlined earlier, and shows that the universal ratio in question is 1.3361 to 4 decimal place accuracy.

The pressure of any dilute gas can thus be used to *define* a universal 'gas temperature,' T, via the relation

$$\frac{T}{T_{ice}} = \frac{p}{p_{ice}}. \tag{7.11}$$

It follows from (7.11) that $T_{steam}/T_{ice} = 1.3661$, which is one equation for the two quantities T_{steam} and T_{ice}. Another condition can be imposed by dividing the interval $T_{steam} - T_{ice}$ into a convenient number of units, and contact with the Celsius scale is made by taking this number to be 100. Then $T_{ice} = 100/(1.3661 - 1) = 273.15$. In short, gas temperature is nothing more or less than the Celsius temperature with 273.15 added to it; on this new scale the freezing point of water at atmospheric pressure, 0°C, is 273.15K, the K being in honor of Kelvin who understood a deeper meaning, to be explained in the next chapter, of this temperature scale.

The above and other experiments on dilute gases can be summarized in the *universal* equation:

$$p V = n R T. \tag{7.12}$$

Fig. 7.4.
Measurements of the ratio (p_{steam}/p_{ice}) for two different gases.

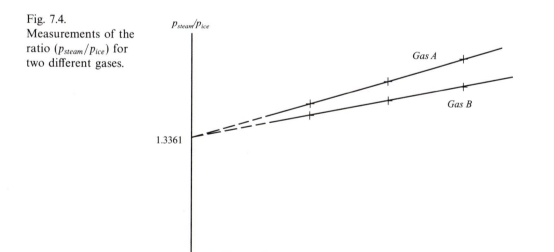

Here p, V, and T are the measured pressure, volume and gas temperature, n is the number of gram molecular weights of the gas,† and R is
the so called universal gas constant, which is approximately 8.3 joules/
K. [The joule is the mks unit of energy: the kinetic energy, $\frac{1}{2}Mv^2$, of
an object of mass $M = 1$kilogram and velocity $v = 1$meter/second is
$\frac{1}{2}$joule. Please verify, incidentally, that the product pV has units of
energy, so that, since n is a pure number with no dimensions, R must
have the units given.]

You may notice a certain similarity between the purely experimental
(7.12), and the purely theoretical (7.9). Before we explore this similarity, which must be telling us something, it is worth examining the
limitations of (7.12), which is called, very grandly, the Ideal Gas Law.
Since all real gases condense into liquids at low enough temperatures,
the formula, which contains no hint of a transition from gas to liquid,
is not true in general. Here, again, the assumption of diluteness is
crucial. Although the formula is only approximately true, it becomes
a better description of reality as gases are rarified and thus moved
further from conditions required for condensation.

Since the restrictions put on (7.9), by examining the approximations
that go into it, are identical to those put on (7.12), by requiring that it
describe reality, what could be more natural than to assume that the
one is the theory of the other. Identifying them for one gram molecular
weight of gas, for which $N = N_{Av}$ and $n = 1$, we get the remarkable
result that, for any dilute gas, twice the average kinetic energy per
molecule is equal to three times the universal gas constant divided
by Avogadro's number and multiplied by the gas temperature. The
coefficient of proportionality, R/N_{Av}, is called Boltzmann's constant,
k_B, and has a numerical value of approximately 1.38×10^{-23}joules/ K.
In terms of this constant, the relation just stated in words is

$$\langle E_{kinetic} \rangle = \tfrac{1}{2}\langle Mv^2 \rangle = \tfrac{3}{2}k_B T \tag{7.13}$$

for any ideal gas. By connecting the mean energy per microscopic constituent of matter with the gas temperature, the molecular viewpoint
breathes life into this otherwise dry as dust concept.

Some comments

I hope that this chapter has brought home the power of the atomic
picture. It allows one to obtain extremely detailed information about

---

† If the mass of the gas in grams is M and its molecular weight is W, n is (M/W). The number of
gram molecular weights is also called the number of 'moles.'

one molecule from experiments involving 10^{23} molecules. Consider the question asked earlier in this chapter: What is the molecular constitution of water? Convert a known weight of water to dilute water vapor at a given gas temperature, T, and measure its pressure, p, and volume, V. Calculate the dimensionless quantity pV/RT, and you know how many gram molecular weights, or moles, of water are contained in the known weight. In this way, one mole of water would be found to be approximately 18 grams. From the already given combining weights of hydrogen and oxygen, we conclude that the molecule of water consists of two atoms of hydrogen and one of oxygen!

It is worth emphasizing that the formula (7.13) is only true for an ideal gas, i.e. one that obeys the ideal gas law (7.12). We shall be considering gas temperature in a more general way in the next chapter. The basic idea that this temperature measures molecular agitation is general. What happens when a red hot pin is dropped into a bucket of water? The pin starts out being much hotter than the water, i.e. its atomic constituents on the average have much more kinetic energy per molecular unit. When a new equilibrium situation is reached, pin and water have a common temperature only very slightly higher than the original temperature of the water because the excess energy of the (iron and carbon) atoms in the pin has been distributed over most of the many water molecules, raising their average kinetic energy by only a little.†

Equation (7.13) also shows that the natural way of measuring temperature is in units of energy. Degrees Celsius or Kelvin are a sort of historical relic. In any microscopic physical context, temperature always occurs in the combination $k_B T$, Boltzmann's constant being the bridge between the natural and the historical units. Indeed one could easily make a mercury-in-glass thermometer that reads temperature in convenient units of 10^{-23} joules. In these units – which I might call Boltzmanns, B – the ice point would be $273.15 \times 1.38 \approx 377$B, and the steam point would be $373.15 \times 1.38 \approx 515$B. At 100B the average kinetic energy of 10^{23} molecules of an ideal gas, through (7.13), is then exactly 150 joules. I hasten to add that this perfectly logical and convenient temperature scale is not in practical use, perhaps because of the weight of historical precedent. In theoretical physics, however, one often finds $k_B T$ replaced by T, which means that temperature is being given in joules. I shall start doing this in chapter 9. First, however, we shall explore the deeper meaning of gas temperature.

---

† In the early (extremely non-equilibrium) stages of this process, some water molecules will have escaped as steam.

Units

In the beginning of Chapter 6, we mentioned the meter, abbreviated m, the second, s, and the kilogram, kg, as the scientific units for length, time, and mass, respectively. Since then, we have encountered the quantities force, energy, pressure, and temperature, all of which have to be specified in units. In this section, you will find a summary of the scientific units – some of them mentioned at various places above, but here pulled together for convenient reference – for mechanical and thermal quantities.

Mechanical units are all expressible in terms of length, time, and mass, but they have been given names of their own.

Quantity	Name	Symbol	Units
force	newton	N	$kg \cdot m/s^2$
energy	joule	J	$kg \cdot m^2/s^2$
pressure	pascal	Pa	$kg/ m \cdot s^2$

The second two can also be expressed in terms of the first. Thus a joule is a newton multiplied by a meter, and a pascal is a newton divided by a square meter.

We have encountered temperature measured in kelvin, K, or, when 273.15 is subtracted from this temperature, degrees Celsius, °C. To convert Kelvin temperature to the temperature in energy units, one must multiply by Boltzmann's constant of magnitude approximately 1.38×10^{-23} joules/kelvin.

We have also encountered the concept of the mole, which is a quantity of matter containing Avogadro's number, approximately 6.02×10^{23}, of molecules.

Finally, in the department of rarely useful information, here are the accepted usages for powers of 10:

peta: 10^{15}, tera: 10^{12}, giga: 10^9, mega: 10^6, kilo: 10^3, hecto: 10^2, deca: 10, deci: 10^{-1}, centi: 10^{-2}, milli: 10^{-3}, micro: 10^{-6}, nano: 10^{-9}, pico: 10^{-12}, femto: 10^{-15}, atto: 10^{-18}.

Thus, a kilogram is 10^3 grams, a micrometer is 10^{-6} meters, and a femtosecond is 10^{-15} seconds.

A hybrid unit for energy, which you may have heard of in connection with electrical energy, is the kilowatthour. The watt is another name for one joule per second. It is thus a unit for specifying rate of change of energy, which has the technical name power. (A 75 watt light bulb converts 75 joules of electrical energy per second into heat and light.) A watthour is a unit of energy. Since a watt multiplied by a second is a joule, a watthour is $60 \times 60 = 3\,600$ joules, and a kilowatthour is a thousand times as much, namely 3.6×10^6 joules.

Solved problems

(1) A hydrocarbon, i.e. a molecule made of only hydrogen and carbon, is known to have a density (mass per unit volume) 2.45 times that of oxygen, when both are dilute gases and at the same temperature and pressure. It is also known that 100 grams of the hydrocarbon contain 7.75 grams of hydrogen. What is its molecular formula? Take it as known that ordinary gaseous oxygen is composed of O_2 molecules. The atomic weights of hydrogen, carbon, and oxygen are (approximately) 1, 12, and 16 respectively.

Solution: This is another example of the power of the atomic hypothesis, showing how observations on a large scale can be made to yield precise insight into the molecular constitution of matter.

The ideal gas law, (7.12), when interpreted microscopically using (7.9) and (7.13), implies that *every* dilute gas at the same temperature and pressure has the *same* number of molecules per unit volume. From this it follows that the molecule of the hydrocarbon in question is 2.45 times as massive as the oxygen molecule! Let the molecular formula be $C_x H_y$. Then, from the just determined ratio of masses, the known molecular formula, O_2, of oxygen, and the known atomic weights, we have

$$12x + y = 2.45 \times 16 \times 2.$$

Furthermore, from the given mass fraction of hydrogen, we have

$$\frac{y}{12x + y} = \frac{7.75}{100}.$$

The solution of this pair of equations is $y = 6.08$, and $x = 6.03$. To the accuracy of the atomic weights used, this is $x = y = 6$. The hydrocarbon is $C_6 H_6$, called benzene.

(2) If 3 moles of a monatomic ideal gas at 20°C are mixed with 2 moles of another monatomic ideal gas at 30°C, and the mixture is thermally isolated and allowed to reach equilibrium, what is the final temperature? [The restriction to monatomic gases is needed because molecular gases have energy associated with the internal structure of molecules. In a dilute monatomic gas, the only energy is kinetic energy.]

Solution: The final temperature is:

$$\frac{3 \times 20 + 2 \times 30}{3 + 2} = 24°C.$$

The reason is that the energy of n moles of a monatomic ideal gas is $\frac{3}{2} n N_{Av} k_B T$, where N_{Av} is Avogadro's number, k_B is Boltzmann's constant, and T is the temperature in kelvin. When the gases are mixed, and thermally isolated, the total energy is conserved. This leads to the formula displayed above. [One can work directly with degrees Celsius, because the additive constant of 273 cancels out of both sides. Check this for yourself.]

(3) A 10 liter container contains 8 grams of oxygen gas at 20°C.

    (i) How many O_2 molecules are there in the container? What is the average volume per molecule? Is the gas dilute?

    (ii) What is the average kinetic energy of an O_2 molecule?

    (iii) What is the root-mean-square (rms) velocity of an O_2 molecule? [v_{rms} is defined by $v_{rms}^2 = \langle v^2 \rangle$]

    (iv) What is the pressure of the gas in the container?

        The container has a piston on top allowing the gas to be compressed. It is compressed to a new volume of 5 liters, still at 20°C.

    (v) What is the new pressure of the gas?

    (vi) What is the change in thermal (kinetic) energy of the gas?

Solution: Since a mole of O_2 gas has a mass of 32 grams, 8 grams of oxygen is 0.25 moles, which corresponds to 1.5×10^{23} molecules in the 10 liter volume. The volume per molecule is thus 10 liters$/1.5 \times 10^{23} = 6.7 \times 10^{-23} 1 = 6.7 \times 10^{-20} cm^3$. The typical volume of a molecule is $(10^{-8} cm)^3 = 10^{-24} cm^3$, which is about 1000 times less than the volume per molecule, so the gas is dilute. The average kinetic energy, E_{kin} , of a molecule at temperature T is $3k_B T/2$, and since $T = 293K$, this gives an average kinetic energy of $6.07 \times 10^{-21} J$. From this we can calculate the root mean square velocity, since $E_{kin} = \frac{1}{2}mv_{rms}^2$. The mass of a single O_2 molecule is $32g/6 \times 10^{23} = 5.33 \times 10^{-26} kg$. This gives us $v_{rms} = 476$ m/s. Finally, the pressure can be obtained from the ideal gas law, $pV = nRT$, using $n=0.25$ moles, $V = 10$ liters $= 0.01 m^3$, $T=293K$ and $R=8.31 JK^{-1} mol^{-1}$. We find that $p = 60,800$ Pa.

If we compress the gas to half its volume at constant temperature, then the kinetic energy of the gas is unchanged, since this depends only upon temperature. In the ideal gas law, the right hand side is unchanged, so halving the volume must double the pressure, giving a final pressure of approximately 122 kilopascals.

(4) Here is a description of a realistic version of Agnes Pockels's experiment. We shall imagine spreading oleic acid – an organic, fatty acid which we shall call 'oil' – on the surface of water, measuring the area of the oil film, and thereby deducing Avogadro's number. In order to get a volume of oil small enough that the film will only partially cover the surface of water in large cookie sheet, the oil is diluted with methyl alcohol (methanol). One puts 5 cm^3 of oleic acid into a graduated cylinder, adds 95 cm^3 of pure methanol, and allows these to mix. One then takes 5 cm^3 of the mixture and mixes it with a further 95 cm^3 of alcohol. In this way one obtains a dilution of 400. One now lightly dusts the water surface with chalk dust by tapping an eraser. A drop of the diluted mixture is dripped on to the water surface, whereupon the alcohol mixes with the water leaving a circular film of oil – made easily visible by the chalk dust now coating the water but not the oil – on the surface. The diameter of the film is

measured with a ruler. The only other experimental observation needed is the volume of a drop, which is obtained by measuring how many drops are needed to fill a small graduated cylinder.

Oleic acid has the chemical structure $(CH_3)(CH_2)_7CH=CH(CH_2)_7$ COOH, showing a hydrophilic (water-loving) COOH group and a hydrophobic (water-hating) hydrocarbon tail. Assume that the molecules form a layer one molecule thick of prisms $l \times l \times 7l$ sticking out of the water and packed close in a square arrangement. The density of oleic acid is 890 kg/(m³).

In an experiment, the film diameter for a single drop was found to be 9.4 cm, and 97 drops were found to occupy 1 cm³. Estimate Avogadro's number N_{Av}.

Solution: First, let's convert the data into usable form. The area of the film A is $\pi d^2/4$, where d is the diameter. This comes out to be $A = 69.4$ cm², which is 6.94×10^{-3} m². The volume of one drop V_{drop} is $(1/97)$ cm³, which is 1.03×10^{-8} m³.

Using the atomic weights of carbon (12), hydrogen (1), and oxygen (16), and noting that the molecular formula has 18 carbon atoms, 34 hydrogen atoms, and 2 oxygen atoms, one works out the molecular weight of oleic acid to be 282. This is thus the number of grams of oil that contain Avogadro's number of molecules. From the given density, this mass (0.282 kgm) occupies a volume $V_m = 0.282/890 = 3.2 \times 10^{-4}$ m³. To get N_{Av} from this volume, we must divide by the volume of the film and multiply by the number of molecules in the film. The volume of the film is $(1/400)$ of the volume of the drop – because of the dilution with alcohol. From the model, if A is the area of the film, A/l^2 is the number of molecules, and the volume of oil is $A \times 7l$.

One thus has two equations for the unknowns l and N_{Av}, namely,

$$l = \frac{V_{drop}}{2800A}, \text{ and } N_{Av} = \frac{400V_m}{V_{drop}} \frac{A}{l^2}. \qquad (7.14)$$

Entering the numerical values for V_m, V_{drop}, and A, one finds $l \approx 0.53 \times 10^{-9}$ m, i.e 0.53 nm, and $N_{Av} \approx 3 \times 10^{23}$ – in astonishing, and probably slightly fortuitous, agreement with the correct 6.02×10^{23}.

8

Disorder, entropy, energy, and temperature

Let me count the ways
Elizabeth Barrett Browning

The most subtle of the concepts that surfaced in the last chapter is *equilibrium.* Although the word suggests something unchanging in time, the molecular viewpoint has offered a closer look and shown temperature to be the average effect of molecular agitation. At first sight, it seems hard to say anything, let alone anything quantitative, about such chaotic motions. Yet, paradoxically, their very disorder provides a foundation upon which to build a microscopic† theory of heat.

How does a dilute gas reach equilibrium? Contemplate a system in a thermally insulating rigid container, so that there is no transfer of energy or matter between the inside and the outside. As time passes, the energy the gas started with is being exchanged between the molecules in the system through random collisions. Finally, a state is reached which is unchanging as far as macroscopic observations are concerned. In this (equilibrium) state each molecule is still engaged in a complicated dance. To calculate the motion of any one molecule, one would eventually need to calculate the motion of all the molecules, because collisions between more and more of them, in addition to collisions with container walls, are the cause of the random motion of any one. Such a calculation, in addition to being impossible, even for the largest modern computer, would be futile: the details of the motion depend on the precise positions and velocities of all the molecules at

† The words microscopic and macroscopic were mentioned in Chapter 3 (p. 26), and the distinction made there between fine and coarse observation is at the root of the meanings of these words in the context of physics. 'Microscopic' implies a description in terms of the basic constituents of matter – molecules, atoms, or sub-atomic particles – even though such objects are not visible through an optical microscope. 'Macroscopic' refers to a description of many millions of atoms or molecules – such as would occur in a quantity of matter visible to the naked eye – in terms of quantities such as volume, pressure, and temperature which can be given meaning without reference to an atomic picture. A macroscopic description need not necessarily have a microscopic picture underlying it; when it does, it deals with averages of microscopic quantities.

some earlier time, and we lack this richness of knowledge. A more modest goal, a more sensible one (since it turns the limited information available into an advantage), and one suggested by the earlier parts of this book, is to seek the *probability* that in equilibrium a molecule is in one of its possible states of motion.†

As in earlier chapters, calculating probabilities will require counting possibilities. This is relatively simple for a *dilute* gas because the rarely colliding molecules can be treated as independent in the long intervals between collisions, so that we can think of the gas as a system of 'weakly interacting' molecules. More generally, the counting may be done straightforwardly for any system made up of a collection of weakly interacting identical 'subsystems,' and we can minimize unnecessary writing by using the subscript i to label the states of motion of one such subsystem.‡

What is the probability p_i that in equilibrium a molecule selected at random has the state of motion specified by the index i? This – the basic question of *equilibrium statistical mechanics* – is answered in this chapter by constructing the probability distribution that solves the problem.

The word probability will be prompting you to identify a repeatable random event. The experiment in question has to be imagined; it is a 'thought experiment.' For a dilute gas, one is to imagine observing how a molecule chosen at random is moving when the gas as a whole is in equilibrium. Now, there are at least two ways in which the probabilities we are interested in could be obtained. One could contemplate following a single molecule in its wanderings about the container and recording its state of motion at times separated by intervals during which many collisions have occurred. From a long series of such observations one could calculate the fraction of events corresponding to a particular state of motion, as an estimate of its probability. Alternatively, one could imagine a measurement at some instant of time of the states of motion of all the molecules in a macroscopic sample of gas. From this information, the fraction of molecules in each possible state of motion could be determined – another estimate of the probabilities.

It is reasonable to suppose that if a macroscopic system does

† The state of motion of a molecule is determined by its position and velocity vectors. Since position and velocity can be changed by arbitrarily small amounts, there are an infinite number of states of motion for a molecule. How the counting of states is to be done in this case will be explained in the last section of this chapter. Meanwhile, I shall ask you to take on faith that no irreparable sin is committed by assuming a finite number of discrete states.

‡ See solved problem (4) at the end of this chapter for an example in which i does not mean position and velocity.

reach equilibrium then in equilibrium these two ways of assigning probabilities to the states of motion of a subsystem will give identical results. We shall make this assumption where necessary in this chapter, i.e. we shall not distinguish between an average over time for a single subsystem and an average over many subsystems at the same time.

A first guess might be that in equilibrium a subsystem picked at random in either of the two ways just discussed is equally likely to be in any of its possible states of motion, but this thought is flawed by a disregard of the fact that an isolated system has a fixed total amount of energy. Since collisions, no matter how random, do not affect the total energy, some subsystems can have more energy than the average only if others have less. A more reasonable assumption would be that each weakly interacting sub-part of a system in equilibrium – i.e. for an ideal gas each molecule – explores all its possible states of motion, but in such a way that its average energy is fixed. To implement this idea, we need a way of characterizing the randomness of a distribution, which brings us to a concept that deserves a section and heading of its own.

Entropy

The term entropy was coined by Clausius in the context of a macroscopic theory of heat, now called thermodynamics, in which molecules and their motion play no role. In thermodynamics, entropy has a meaning which generations of science and engineering students have found slightly mysterious. For most of us, the mystery was only dispelled when we encountered the same word in the present molecular, microscopic, statistical context. You are about to have the good fortune of *first* meeting entropy in this guise, where it is associated with disorder, and its meaning has entered popular usage. If you have heard the word before, you probably would not be surprised to hear an increase of disorder described as an increase of entropy; if you have not, the idea is roughly contained in the statement that a house of cards collapsed has *more* entropy than the same house of cards standing.

Useful concepts in physics have a quantitative side to them, and entropy is no exception. Can one associate a number with a distribution, which number is to describe the 'randomness' of the distribution? This may seem too vague a question to have any answer at all, let alone an answer that is to all intents and purposes unique. And yet, by making certain reasonable requirements, we shall find such an answer. It is useful to contemplate the wanderings of a sub-part of a macroscopic system, one molecule in a dilute gas for example, through its possible states of motion. It is natural to associate a decrease in order with an increase in the number of states of motion that are explored. For a dilute gas, this would be the state of affairs in which, as a result of collisions, a given molecule successively has as wide a range of positions and velocities as possible. When the gas is liquified, each molecule is more constrained and passes through fewer of its possible states of motion. Thus the number of allowed states would seem like a good measure of the idea we are trying to quantify. There are two problems with this. One, it is not unique: any function† of the accessible number of states which increases when that number increases will do. Two, we have not allowed for the fact that some accessible states may be more probable than others. Let us, at first, ignore the second difficulty, and consider only distributions corresponding to I equally likely alternatives. Then, the required function $f(I)$ turns out to be fixed by the requirement of *additivity*, which is best explained by a simple example.

Suppose that a part of a system has three equally likely states of motion, call them a, b, c. We want the entropy of this subsystem to be the function we are seeking evaluated at the input 3, namely $f(3)$. Suppose that another subsystem has four equally likely states of motion, 1, 2, 3, 4, or an entropy of $f(4)$. Now, the two considered together have 3×4 ($a1$, $a2$, $\dots c4$) or 12 equally likely states, if (and only if) the two subsystems are weakly interacting. [Otherwise there might, for example, be an enhanced likelihood of the second subsystem being in state 1 when the first was in a, making $a1$ more likely than $a2$.] It seems reasonable to require that the entropy of two weakly interacting subsystems be equal to the sum of their entropies, because one would not want the measure of randomness in one subsystem to depend on the state of affairs in some other, possibly remote, subsystem.‡ The requirement leads to the condition that the unknown function f have the property $f(3 \times 4) = f(3) + f(4)$. More generally, we could replace 3 and 4 by any integers. The function that has this property is, as we know from Chapter 5, the logarithm. Logarithms to any base greater than 1 will do, and this ambiguity, which corresponds to multiplying $f(I)$ by an arbitrary positive constant, is the only one

---

† Recall the definition of Chapter 5, where a function was defined as a number that depends on another number. Below, we shall use the word in a slightly more general way, to denote a number that is determined by many numbers.

‡ In this respect we are requiring entropy to be like energy: the energy of two remote subsystems is equal to the sum of their separate energies.

left. There are various conventions. We shall do something slightly
unconventional in the context of statistical mechanics, follow the
practice in what is called 'information theory,' and take e as the base,
i.e., identify $f(I) = \ln I$. To put the matter in a slightly different way,
if the equally likely states of motion are labeled with the index i, so
that the the normalized probability distribution is given by

$$p_i = \frac{1}{I}, \quad i = 1, 2, \ldots I, \tag{8.1}$$

we have argued that the entropy to be assigned to the subsystem is
$\ln I$.

Now, i labels a microscopic state of a subsystem, and I is thus
the number of equally likely 'microstates' in a macroscopic description. <u>The entropy of a subsystem is thus the natural logarithm of the
number of equally likely microstates corresponding to a given
'macrostate.'</u>

Now we must assign an entropy to a distribution with unequal
entries. We need to do this because, according to the program proposed
at the end of the last section, we shall be seeking the probability
distribution for equilibrium, in which every state of motion i is *not*
equally likely. Therefore let us now assume only that the probabilities
p_i are a distribution in the sense of Chapter 3. Let $S(p_1, p_2, \ldots p_I)$
be the entropy to be associated with the distribution. This general
definition must reduce to the previous one when the p_is are all equal.
In particular, we must have $S(\frac{1}{3}, \frac{1}{3}, \frac{1}{3}) = f(3) = \ln 3$. What number
should we associate with $S(\frac{1}{3}, \frac{2}{3})$?

This question can be addressed in two different ways. The first
method, given in this paragraph, has the advantage of dealing explicitly
with equally likely alternatives but the drawback of requiring Stirling's
formula (Chapter 5) for the logarithm of the factorial of a large
number. Contemplate a very large number N of weakly interacting
identical subsytems each of which can be in one of two possible states,
and suppose that in equilibrium the probabilities are $\frac{1}{3}$ and $\frac{2}{3}$. The
mean numbers of subsystems in the two states will then be $N/3$ and
$2N/3$. [Since N is a large number, $N \gg \sqrt{N}$, and fluctuations about
these averages are negligible for reasons explained in Chapter 3.] As
time progresses these average numbers will remain constant, but *which*
of the subsystems are in one state or the other will change. Equally
likely microstates thus correspond to the different ways in which the
N subsystems can be divided into those in state 1 and those in state
2. We can calculate this number, call it C, using methods learnt in

Chapter 2. It is the combinatorial coefficient

$$C = \frac{N!}{(N/3)!(2N/3)!}.$$

(8.2)

The logarithm of this number of equally likely ways should thus be interpreted as the entropy of N subsystems, or N times the entropy per subsystem. For very large N, $N/3$, and $2N/3$ Stirling's formula may be used in the form $\ln(N!) \sim N \ln N - N$. It then follows from (8.2) that

$$\ln C \sim N \ln N - \frac{N}{3} \ln \frac{N}{3} - \frac{2N}{3} \ln \frac{2N}{3} - N + \frac{N}{3} + \frac{2N}{3}, \quad (8.3)$$

with the right hand side being a better approximation to the left as N is made larger. Now, note that the last three terms add up to zero. Further, since $\ln(2N/3) = \ln N + \ln(2/3)$ and $\ln(N/3) = \ln N + \ln(1/3)$, all terms proportional to $N \ln N$ cancel out, leaving finally

$$\ln C = -N\left[\tfrac{1}{3} \ln \tfrac{1}{3} + \tfrac{2}{3} \ln \tfrac{2}{3}\right] = NS(\tfrac{1}{3}, \tfrac{2}{3}).$$

(8.4)

Since there is nothing special about the distribution $\tfrac{1}{3}, \tfrac{2}{3}$ we have the result

$$S(p_1, p_2) = -[p_1 \ln p_1 + p_2 \ln p_2]$$

(8.5)

for any two-state subsystem.

This method may be generalized in a straightforward way to apply to weakly interacting subsystems with I states and associated probabilities $p_1, p_2, \ldots p_I$, Consider a large number N of such subsystems, and note that on the average there will be Np_i systems in state i, where $i = 1, 2, \ldots I$. The number of ways in which N identical objects may be divided into bins containing Np_1, Np_2, etc. is

$$C = \frac{N!}{(Np_1)! \times (Np_2)! \times \ldots (Np_I)!},$$

(8.6)

the denominator here, as in (8.2), correcting for the re-arrangement of identical subsystems in a given bin.

Using Stirling's formula to work out the logarithm of (8.6) as in the two-state case one obtains for the entropy per subsystem

$$S(p_1, p_2, \ldots p_I) = -[p_1 \ln p_1 + p_2 \ln p_2 + \ldots p_I \ln p_I].$$

(8.7)

Please check that if the p_is are now taken to be equal – whereupon each must equal $\tfrac{1}{I}$ to give a normalized distribution – (8.7) reduces to $\ln I$.

Another derivation

Equation (8.7) is sufficiently important to justify thinking it out in another way. This second method does not require the magic of Stirling's formula, but does employ rather subtle reasoning. Consider again a two-state subsystem with probabilities $\frac{1}{3}$ and $\frac{2}{3}$. Think of the distribution $\frac{1}{3}, \frac{2}{3}$ as describing a system that fluctuates between two states of motion, on the average spending one third of its time in the first state, and two thirds in the second. Suppose we thought of the time spent in the second state as being equally divided between two 'imaginary' distinct states; then we would be imagining the motion as being equally divided between a total of three distinct states. However, the entropy associated with an equal three way division is too large, because for two thirds of the time there is in fact no real mixing of states. A reasonable way to correct for this overcounting is to subtract from $S(\frac{1}{3}, \frac{1}{3}, \frac{1}{3})$, i.e. the entropy associated with a complete mixing between three states, two-thirds of the entropy, $S(\frac{1}{2}, \frac{1}{2})$, associated with a complete mixing between the two imaginary states. The mathematical requirement is thus:

$$S(\tfrac{1}{3}, \tfrac{2}{3}) \;=\; S(\tfrac{1}{3}, \tfrac{1}{3}, \tfrac{1}{3}) - \tfrac{2}{3} S(\tfrac{1}{2}, \tfrac{1}{2}). \tag{8.8}$$

The entropies on the right hand side of this equation are for equally likely states, and have been previously taken to be the logarithms of the number of states. Making these substitutions, we obtain

$$
\begin{aligned}
S(\tfrac{1}{3}, \tfrac{2}{3}) &= \ln 3 \;-\; \tfrac{2}{3} \ln 2 \\
&= \tfrac{1}{3} \ln 3 \;+\; \tfrac{2}{3} \ln \tfrac{3}{2} \\
&= -[\tfrac{1}{3} \ln \tfrac{1}{3} + \tfrac{2}{3} \ln \tfrac{2}{3}].
\end{aligned}
\tag{8.9}
$$

The purpose of the algebra leading from the first to the second line, which you should please check, is to write the entropy for unequal probabilities in terms of them. Note that the result in (8.9) is equivalent to (8.4); (8.5) follows from it as before.† Equation (8.7) may be obtained by generalization of the partitioning requirement (8.8). This is shown in solved problem (1) at the end of this chapter.

---

† Note that the properties of large numbers came into the previous derivation but not into this one. In fact there is some ambiguity about what one means by entropy unless very large numbers are implicit. This point will be clarified in Chapter 10.

Some properties of entropy

Before discussing the mathematical properties of the function we have introduced to characterize disorder, here is an example that may clarify the quantitative connection between increasing disorder and increasing entropy. Consider an ideal gas in a container equipped with a partition that divides the enclosed volume into two equal parts. Suppose to begin with that all the molecules are to the left of the partition. If the partition is removed, and the gas allowed to fill the whole space, the disorder and, correspondingly, the entropy of the system will increase. By how much? The ideal gas, according to our previous discussion, is made up of weakly interacting molecules. Since a given molecule is equally likely to be – i.e. to have a position vector – anywhere within its region of confinement, and since the space filled by the allowed position vectors for one molecule doubles when the partition is removed, the number of available states of motion is also doubled. Thus the increase in the entropy is $\ln 2X - \ln X$, where $\ln X$ is the entropy of one molecule on the left of the partition. If there are N molecules in the gas, then, because in the dilute gas in question the molecules individually are weakly interacting subsystems, and because we have required that the entropy for such subsystems be additive, the total increase in entropy is $N \ln 2$. [Note that X has canceled out because $\ln a - \ln b = \ln(a/b)$.]

Now let's understand the meaning of (8.5) which must seem rather inscrutable, but will lose its mystery as we work with it. The first thing that needs to be demystified is the function $s(x) = x \ln(1/x) = -x \ln x$ for x between 0 and 1, because S is a sum of terms of this form, and because each of the ps is a member of a distribution (cf. Chapter 3) and therefore lies in this range. One of the curves in Figure 8.1 is a plot of this $s(x)$. Note that it is zero at the end points of the range, is never negative, and has a single maximum of about 0.37 when x is about 0.37. For what follows it will be useful to understand that one can work out the position of this maximum without going to the trouble of plotting the whole curve. Let us work out the change in s when x is changed by a little. As a first step it is useful to do this for the logarithm. Using results obtained in Chapter 5, and the notation introduced in Chapter 6 for a small change, we have

$$\ln(x + \Delta x) = \ln[x \times (1 + \Delta x/x)] = \ln x + \ln(1 + \Delta x/x)$$
$$= \ln x + \Delta x/x + \ldots \tag{8.10}$$

where the $\ldots$ indicates terms of higher powers of Δx.

Equation (8.10) leads to the result

$$\Delta s = s(x + \Delta x) - s(x)$$
$$= -(x + \Delta x)\ln(x + \Delta x) - x\ln x \tag{8.11}$$
$$= -\Delta x[\ln x + 1] + \dots,$$

where we have carefully kept all terms proportional to Δx

At its maximum, $s(x)$ is level – as are all smooth functions at their maxima or minima – and therefore does not change when x is changed by the small amount Δx. From (8.11) we notice that this condition obtains when the quantity in square brackets is zero, i.e. when $\ln x = -1$. This, as Chapter 5 has taught you, means $x = 1/e$, or, substituting the numerical value of e, $x = 0.367\dots$ You will also readily verify that $s(e^{-1}) = e^{-1}$, so that the little mathematics of Chapter 5 has allowed us to work out the position and numerical value of the maximum of $s(x)$ directly.

This method of identifying the level points, i.e. the extrema, of a function by comparing small differences will serve us well in what follows. [The method is, in fact, that of the differential calculus – though the name is less important than the reasoning, for the details of which Chapter 5 has, I hope, completely prepared you.]

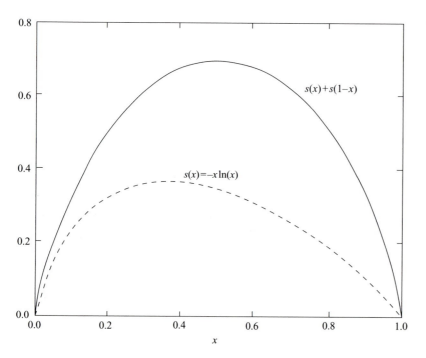

Fig. 8.1. Graphs of $s(x) = -x\ln x$ and of $S(x, 1-x) = s(x) + s(1-x)$

The other curve plotted in Fig. 8.1 is $s(x) + s(1-x)$. Note, via (8.5), that this is the entropy of a two-state system with probabilities x and $1-x$, i.e., the graph is of $S(x, 1-x)$. What do we learn from the curve? At each extremity, $x = 1$ or $x = 0$, $S(1,0) = S(0,1) = 0$. These two end points correspond to situations with no mixing at all; the system may have two possible states, but it is always in one or the other; the entropy associated with these two limits of perfect order is zero. In between, the entropy rises symmetrically, reaching a maximum for $x = 1 - x = \frac{1}{2}$. At this point, which corresponds to perfect mixing between two states, the entropy is $\ln 2$, which agrees with $S = \ln I$, $I = 2$, for equally probable states.

Some of the properties of the entropy of a two-state system are completely general and apply to (8.7):

(i) For complete mixing between I states – all ps equal to $\frac{1}{I}$ – the entropy is $\ln I$.

(ii) The entropy, being a sum of non-negative terms, is never negative.

(iii) When there is no mixing, i.e. when the distribution corresponds to a certainty – all but one of the ps equal to 0 and the remaining p equal to 1 – the entropy is <u>zero</u>, because it is a sum of $I - 1$ terms $s(0)$, each of them zero, and a single $s(1)$, also zero.

We shall see shortly that (i) gives the maximum entropy possible for a system with I states. It will then become clear that the quantity defined in (8.7) has all the properties we want for something that measures mixing or randomness in the sense discussed here: it is zero when a subsystem is confined to any one of its possible states of motion; it reaches its maximum when the system has an equal probability of being in each of its states.

Maximizing entropy: minimizing order

We come now to what is perhaps the strangest calculation in the standard curriculum of physics. We seek a probability distribution (i.e., a set of numbers, each between zero and one, whose sum is unity) which maximizes the entropy (a single number), subject only to the constraint of conservation of energy. The 'outcomes,' in the sense of Chapters 1–4, that the probabilities refer to are those discussed in the earlier parts of this chapter: states of motion of sub-parts of a large dynamical system. The calculation is strange because so much is obtained for what, on the face of it, seems like so little. It is also strange in that, although the complicated dance discussed at the start of this chapter is central to it, time does not enter the calculation at all. The result is a unique time-independent distribution that, we shall claim, describes equilibrium and gives a very precise and insightful meaning to the concept of temperature.

The least problematical way of understanding what this distribution means may be to recognize that we are considering a system that consists of many, many weakly interacting subsystems. When the system is in equilibrium, each of its subsystems is fluctuating. However, the number of subsystems in a particular state of motion is on the average independent of time, and it is this constant average fractional occupancy that is to be calculated, using the idea that in equilibrium a system is as mixed-up as it can be.

Although time is nowhere explicit in the calculation, it does enter the most basic physical assumption underlying the method. This is that only those states of motion which can be reached in the time of observation are to be included. An example will help. Consider the equilibrium of some helium gas in an ordinary rubber balloon. Only position vectors *within* the balloon would normally be allowed for the helium atoms. However, if the equilibrium in question is several days in the future, when most of the gas will have diffused through the balloon's surface, leaving behind a child's disappointment drooping on its bedside string, this assumption would be incorrect. For any system that is not absolutely stable, restrictions of this kind – not always as trivial to implement as in the simple example – are implicit.

As a warm-up exercise, let us first ignore conservation of energy and ask for the distribution that gives the maximum entropy pure and simple. Start with (8.7) which, using the notation for a sum introduced in Chapter 4, may be written as

$$S = -\sum_{i=1}^{I} p_i \ln p_i \qquad (8.12)$$

where, since the *p*s make up a distribution, we have

$$\sum_{i=1}^{I} p_i = 1. \qquad (8.13)$$

Now, contemplate two distributions, p_i^0 and $p_i = p_i^0 + \Delta p_i$, with $i = 1, 2, \ldots, I$, where the elements of the second distribution are, as the notation is supposed to indicate, only slightly different from the corresponding elements of the first. How does S change when the distribution is thus slightly changed? The answer is given by (8.11) for each term shown explicitly in (8.7) and symbolically in (8.12). Thus

$$\Delta S = -\sum_{i=1}^{I} \Delta p_i [\ln(p_i^0) + 1] + \dots \tag{8.14}$$

Using the same reasoning as in the last section, an extremum (maximum or minimum) of S occurs when the change in S is zero to linear order (i.e. when the sum displayed on the right of (8.14) is zero) for all possible changes in the distribution.

Now comes a tricky point. In going from one distribution to another, one is not free to change just one of the p_is, because that would violate the normalization requirement, (8.13). The smallest change, consistent with this requirement, is to increase one of the elements by a small amount, and reduce another by the *same* amount. Contemplate a change in the probabilities of states 1 and 2, such that $\Delta p_1 = -\Delta p_2$, with none of the other probabilities changed. For this change, (8.14) reads

$$\Delta S = -\Delta p_1 [\ln(p_1^0) + 1 - \ln(p_2^0) - 1]. \tag{8.15}$$

If the distribution p_i^0 corresponds to an extremum of S, the right hand side of (8.14) must be zero, which means that the square bracket must be zero, the logarithms must be equal, and, consequently, that $p_1^0 = p_2^0$. Now, the labels 1 and 2 in this exercise could have been *any* pair. We conclude that the distribution that gives S its extreme value has all entries equal, and thus each entry equal to $\frac{1}{I}$. The entropy S is then, as we know, equal to $\ln I$. We can also conclude that this extremum is a maximum – because S has only one level point, is never negative, is smooth, and is zero for the special cases discussed in item (ii) at the end of the last section. (Try to draw a curve that has these properties and anything but a single maximum.) The distribution that maximizes the entropy thus does give the maximally mixed condition.

A minor technical remark is worth making here. The cancellation of 1s in (8.15) is not an accident. Quite generally, the expression for ΔS to linear order in the Δps may be written

$$\Delta S = -\sum_{i=1}^{I} \Delta p_i \ln p_i^0 + \dots \tag{8.16}$$

This equation is *identical* to (8.14) because the term dropped is equal to zero. (The following is a good check that you still are on top of the $\sum$ notation. If anything is unclear, write the equations without using this notation, as, e.g., (8.7) was written.) It is zero because p_i^0 and

$p_i = p_i^0 + \Delta p_i$, $i = 1, 2, \ldots I$, are, by hypothesis, both *distributions*, so that the sum of their elements is 1. Thus the sum of the differences in their elements is equal to zero, i.e. $\sum \Delta p_i = 0$. It is also worth noting that $\sum K \Delta p_i = K \times \sum \Delta p_i = 0$, when K is an *i*-independent constant, because the constant, multiplying every term in the sum, can be taken out as an overall multiplicative factor as indicated.

Now, at last, we have in working order all the machinery needed to implement the program mentioned many times, in many forms, throughout this chapter. We add to (8.12) and (8.13) the condition that our subsystem has a fixed average energy, and seek the distribution that maximizes the entropy subject to this condition. If we associate with the state of motion *i* the energy ϵ_i, ('epsilon') the mean energy of the subsystem (cf. Chapter 4) is

$$E = \sum_{i=1}^{I} \epsilon_i p_i. \tag{8.17}$$

(If you are getting tired of the extreme generality, you may wish to keep in mind that for the ideal gas *i* refers to the position and velocity vectors of a single molecule, in which case ϵ_i would be the kinetic energy – half the mass times the velocity squared – of that state of motion.) Because of the extra condition of constancy of energy, the smallest number of probabilities that change in moving away from an extremum of S is three instead of the previous two. As before, let p_i^0 for $i = 1, 2, \ldots I$ be the distribution that maximizes the entropy, and let Δp_i^0 be small deviations. Contemplate, for definiteness, a situation in which only Δp_1, Δp_2, and Δp_3 are the only non-zero deviations. The following three equations must then be obeyed:

$$\Delta p_1 + \Delta p_2 + \Delta p_3 = 0, \tag{8.18}$$

$$\epsilon_1 \Delta p_1 + \epsilon_2 \Delta p_2 + \epsilon_3 \Delta p_3 = 0, \tag{8.19}$$

$$-\Delta p_1 \ln p_1^0 - \Delta p_2 \ln p_2^0 - \Delta p_3 \ln p_3^0 = 0. \tag{8.20}$$

The first of these equations requires that the distribution remain normalized after the contemplated small change; the second, that the energy remain constant as the distribution is changed; and the last, that the entropy remain unchanged to linear order in the small deviations.

What do these equations imply about the distribution p_i^0? A little algebra is needed. The answer, given in (8.26) is more important than the details that follow, which, however, are all within your reach. Multiply (8.18) by ϵ_3 and subtract it from (8.19) to get

$$(\epsilon_1 - \epsilon_3)\Delta p_1 + (\epsilon_2 - \epsilon_3)\Delta p_2 = 0. \tag{8.21}$$

Multiply (8.18) by $\ln p_3^0$ and add it to (8.20) to obtain

$$(\ln p_3^0 - \ln p_1^0)\Delta p_1 + (\ln p_3^0 - \ln p_2^0)\Delta p_2 = 0. \tag{8.22}$$

Since the last two equations must give the same ratio $\Delta p_1/\Delta p_2$, we have:

$$\frac{\epsilon_1 - \epsilon_3}{\epsilon_2 - \epsilon_3} = \frac{\ln p_3^0 - \ln p_1^0}{\ln p_3^0 - \ln p_2^0}, \tag{8.23}$$

which is equivalent to

$$\frac{\epsilon_1 - \epsilon_3}{\ln p_3^0 - \ln p_1^0} = \frac{\epsilon_2 - \epsilon_3}{\ln p_3^0 - \ln p_2^0}. \tag{8.24}$$

Now, the left side of this equation does not depend on the label 2, and the right side does not depend on the label 1. So the equality can only be satisfied if both sides are some function of the state 3. But we could have chosen any state to play the role of 3. Thus, the only way to satisfy (8.24) is to have each side to be equal to a label-independent constant. We shall call this constant $1/\beta$ ('beta'). It then follows that

$$\beta\epsilon_1 + \ln p_1^0 = \beta\epsilon_2 + \ln p_2^0 = \beta\epsilon_3 + \ln p_3^0. \tag{8.25}$$

Again, any three states could have been chosen in making the small deviation from the maximizing distribution. We conclude that the three expressions in (8.25) are not only equal to each other but are equal to the same expression for an arbitrary label $i = 1, 2, \ldots I$, i.e., to a state-independent constant. Call it α ('alpha'). Then $\beta\epsilon_i + \ln p_i^0 = \alpha$, for $i = 1, 2, 3 \ldots, I$. Rearranging terms, and undoing the logarithms by exponentiation, yields the astonishingly simple result

$$p_i^0 = e^{\alpha - \beta\epsilon_i}. \tag{8.26}$$

This is called the Gibbs distribution, after Josiah Willard Gibbs (1839–1903), or the canonical distribution (Gibbs' own rather imposing name for it). Its form is sufficiently simple that one can penetrate some of its secrets easily. We shall be doing exactly that very shortly. First, let us tidy it up a little. Note that the two unknown constants α and β are exactly what is needed to satisfy the constraints of normalization

(8.18) and fixed average energy (8.19). We can immediately fix α by imposing (8.13) on (8.26):

$$1 = \sum_{i=1}^{I} p_i^0 = e^\alpha \sum_{i=1}^{I} e^{-\beta \epsilon_i}. \tag{8.27}$$

Using (8.27) in (8.26), the entries in the Gibbs distribution take the form

$$p_i^0 = \frac{e^{-\beta \epsilon_i}}{Z}, \quad \text{where } Z = \sum_{j=1}^{I} e^{-\beta \epsilon_j}. \tag{8.28}$$

This is now explicitly normalized. [In the definition of Z we have used the index j, which runs over all possible values, in order not to confuse this sum with the particular state i.] The remaining constant, β, is connected to the average energy via (8.17), which can only be worked out in detail for particular situations in which the 'states of motion' i and the associated energies ϵ_i are specified. The ideal gas suggests itself as an interesting avenue for exploring this connection, since we know from the last chapter how to relate the average energy of an ideal gas molecule to the gas temperature introduced there.

Before pushing on, let us step back from the trees and take a look at the woods we have traversed so far in this chapter. A few simple but deep ideas have carried us a long way: *equilibrium*, as the condition in which the parts of a statistical system are 'stirred' as much as possible; *entropy*, as a way of quantifying this, on the face of it, vague thought; *maximization of entropy* subject to the constraint of fixed average energy, as a way of deducing the probability that a subsystem will be found in a particular one of its states of motion. These thoughts have led us to the Gibbs distribution. We have now to verify that the properties we expect of equilibrium are described by this distribution, and also to make connections between it and what we have already learned about dilute gases.

Absolute temperature

If the Gibbs distribution describes equilibrium, it must somehow encompass the concept of temperature. That it does, and very beautifully too, is the topic to be considered next. Let us briefly review what we already know about temperature. In the last chapter we thought of it as 'hotness,' quantified by thermometers. The constant volume gas thermometer was particularly interesting, because we could make a kinetic theory of the 'working substance,' a dilute gas. The theory enabled us to give life to the natural temperature scale of this device, the gas temperature, by showing it to be proportional to the mean kinetic energy per molecule. However, the concept remained tied to rarified gases and to a particular measuring instrument. Here, we have been both more general and more detailed: we have moved from a discussion of the ideal gas to the case of weakly interacting subsystems, without requiring the later to be individual molecules, and we know not only about the mean energy but about the distribution of energies. This distribution contains a parameter, β, which is determined by the mean energy, suggesting a connection between β and temperature. We shall show in what follows that

$$\beta = \frac{1}{k_B T}, \tag{8.29}$$

where T is a temperature scale more general than but numerically identical to the gas temperature: (i) more general than it because any two systems in equilibrium, in contact with each other, and isolated from the rest of the world have the *same* β; (ii) identical to it because if the system is an ideal gas the mean kinetic energy per molecule is $3/2\beta$, or through (8.29) $\frac{3}{2}k_B T$ – the *same* formula deduced in the last chapter (7.13) where T was the empirical gas temperature. The Gibbs distribution thus does indeed feature temperature, via (8.29) with T measured in K (kelvin), as the central attribute of equilibrium.

The remark (i) is in part implicit in the derivation just given of the Gibbs distribution, but also goes beyond it. First, note that the parameter β is common to *all* the identical subsystems which have been allowed to exchange energy via (8.19). Thus, if we were to imagine the system divided into two weakly interacting parts, each part would have the same β. More generally, suppose that we have a different system made up of another collection of identical subsystems with energy states ϵ'_j. Let the total entropy of the two systems be maximized subject to the condition of fixed total energy. This can be done by imagining an 'ensemble' of which each member is a pair of subsystems, one of each kind.† Since the energies of the states of motion of one of these combined subsystems are of the form

$$E_{ij} = \epsilon_i + \epsilon'_j, \tag{8.30}$$

---

† Ensemble as a word to describe a collection of weakly interacting subsystems is, like canonical to describe (8.28), a coinage due to Gibbs. His ornate usages have become standard in physics: the distribution (8.26) or (8.28) is said to describe a 'canonical ensemble.'

the equilibrium probability of finding the unprimed system in the state with energy ϵ_i and the primed system in the state with energy ϵ'_j is, in complete analogy with (8.28), proportional to

$$e^{-\beta E_{ij}} = e^{-\beta[\epsilon_i + \epsilon'_j]} = e^{-\beta \epsilon_i} e^{-\beta \epsilon'_j}. \tag{8.31}$$

But this is just the probability of two independent systems each in equilibrium but with same β. Systems in equilibrium and in equilibrium with each other thus have a common β. The identification (8.29) thus satisfies this intuitive requirement for a reasonable definition of temperature.†

The justification of (ii) requires a complicated calculation which is carried out in the next section. Before embarking on this, it is probably not a bad idea to attempt some further intuitive understanding of (8.29). As a start, let us check that this equation satisfies the physically sensible requirement that its two sides describe quantities with the same dimensions. From (8.25), we see that $1/\beta$ has the dimensions of energy. So, you will recall from (7.13), does $k_B T$. So this check comes out all right.

As a next small step, consider the energy dependence of (8.28) at fixed β. This is illustrated in Fig.8.2. Because the denominator is the

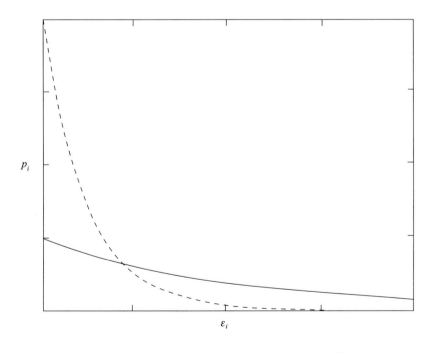

Fig. 8.2. Illustrative graphs of Gibbs distributions. The dashed line shows a large β (low-temperature) distribution, while the solid line shows a small β (high-temperature) distribution.

---

† Another intuitively reasonable property of temperature (that heat flows from high temperature to low temperature) will be demonstrated to hold for (8.29) in the next chapter.

same for every i, the ϵ_i dependence is contained in the numerator which decreases rapidly as ϵ_i increases.† Thus the more energetic a state of motion the less likely it is to be appreciably populated. Correspondingly, as β is decreased, the 'reach' of the exponential increases, and states of higher energy acquire an appreciable probability of occupation. Smaller β thus corresponds to larger average energy, suggesting that a reciprocal relationship between β and temperature is sensible.

It will not have escaped you that these observations are not quantitative. Why, for example, is the constant of proportionality the Boltzmann constant, introduced in the last chapter in connection with dilute gases, and not some other constant? There is no way of avoiding an explicit calculation of the β of an ideal gas described by the Gibbs distribution.

Gas temperature and absolute temperature are identical

The central result to be obtained here is the connection between the average kinetic energy of a molecule in an ideal gas described by the Gibbs distribution and the β of that distribution:

$$< E_{kinetic} >= \frac{3}{2\beta} = \tfrac{3}{2}k_B T, \tag{8.32}$$

where (8.29) has been used in the last equality, so that T is the absolute temperature. The importance of this result is worth repeating: a comparison of (8.32) with (7.13), in which T is the empirical temperature measured by a gas thermometer, shows that the title of this section is true for an ideal gas. It is true in general because according to the reasoning in the last section *any* system in equilibrium with an ideal gas shares its absolute, and thus its gas, temperature.

This section is harder to follow than previous ones, in which some elements of the differential calculus were made up out of whole cloth, because the same thing has to be done here with some elements of the integral calculus. If what follows causes you more than a creative

---

† Note that, β is always being taken to be positive. Negative β would have the feature, extremely counter-intuitive for a system in equilibrium, of making high energy states *more* likely, and, worse, would make it impossible to satisfy the normalization requirement (8.13) whenever states of arbitrarily high excitation are available. The last is the important restriction: when only a band of energy states is allowed the probability distribution can at least temporarily have higher energies being more probable. Such 'negative temperature' distributions occur in the operation of lasers, which are briefly discussed in Chapter 12.

amount of discomfort, it is more important to understand the general drift of the argument and the significance of the result than the details of the proof. However, the application of the general ideas of the earlier parts of this chapter to a particular case is instructive, and the details require nothing that you have not learned in this book.

The first thing that has to be understood is how to sum over the states of motion of a molecule, which, as in Chapter 7 we take to be a structureless point mass. A gas molecule can be anywhere in the containing vessel and have any velocity vector. How is one to accommodate this continuous set of states within the i notation? This turns out to be a non-trivial point. At the purely technical level, we can do the trick by dividing up space into small cubes Δx on a side, and thus of volume $(\Delta x)^3$, and saying that the position of the molecule is specified by saying which box is occupied. Similarly, but more abstractly, we can think of the velocity vectors as filling 'velocity space,' and divide this up into small cubes of 'volume' $(\Delta v)^3$. Then i is specified by which box in ordinary space and which box in velocity space is occupied. In short, we have decided to specify where a molecule is and how fast it is moving only to a certain degree of precision.

Now, the energy of an ideal gas molecule is its kinetic energy $\epsilon_i = \frac{1}{2}Mv^2$, where M is the mass of the molecule. Thus the Gibbs distribution for a molecule is

$$p(\vec{x}, \vec{v}) = \frac{1}{Z} e^{-\frac{1}{2}\beta M v^2}. \tag{8.33}$$

(This is also called the Maxwell distribution, after J. C. Maxwell, who worked it out in the context of the theory of gases.) Note that the state of motion, previously denoted by i, is now labeled by the position and velocity vectors, for which the notation $\vec{x}$ and $\vec{v}$ has been used. A problem now arises. The quantity p is a probability, and thus a pure number. When summed over all boxes, the distribution $p(\vec{x}, \vec{v})$ must be normalized, i.e., give the number 1. However, the the product of an $\vec{x}$ box and a $\vec{v}$ box has the dimensions of (length)^6/(time)^3. If we sum (8.33) over a fine grid without multiplying by the small volume of each box we cannot get a finite number. Let me introduce the symbol ω ('omega') to stand for a small number with the dimensions just given, as this size. [We are here skirting around the edges of quantum mechanics, which specifies a numerical value of ω so small that the sum over boxes becomes independent of their size.] The condition that (8.33) be a probability distribution can now be written as

$$\frac{1}{\omega} \sum_{boxes} (\Delta x)^3 (\Delta v)^3 p(\vec{x}, \vec{v}) = 1. \tag{8.34}$$

The sum over boxes requires a knowledge of the integral calculus to perform in detail. However, two general features can be obtained with little or no work. First, observe that the only $\hat{x}$ dependence comes from the sum over position boxes and gives the total volume of the container – call it V. Second, note that the combination $\frac{1}{2}\beta Mv^2$ is dimensionless – because $\frac{1}{2}Mv^2$ is an energy and β is the reciprocal of an energy. This means that if the velocity box is multiplied by $(\beta M/2)^{3/2}$ it is replaced by the dimensionless quantity $(\Delta v\sqrt{\beta M/2})^3$. From these two simple facts we conclude, using (8.33), that (8.34) has the form

$$\frac{1}{\omega}\sum_{boxes}(\Delta x)^3(\Delta v)^3\frac{e^{-\frac{1}{2}\beta Mv^2}}{Z} = \frac{1}{Z\omega}V\left(\frac{2}{\beta M}\right)^{3/2}K = 1 \quad (8.35)$$

where K is a dimensionless constant given by

$$K = \frac{1}{V}\sum_{boxes}(\Delta x)^3(\Delta v\sqrt{\beta M/2})^3 e^{-\frac{1}{2}\beta Mv^2}, \quad (8.36)$$

whose numerical value can be obtained by using the integral calculus,† but is *not* needed to obtain the quantity we want to calculate:

$$\langle\text{Kinetic energy/molecule}\rangle = \frac{1}{\omega}\sum_{boxes}(\Delta x)^3(\Delta v)^3\frac{1}{2}Mv^2 p(\hat{x},\hat{v}). \quad (8.37)$$

One can sneak up on this in the following tricky way. In (8.35) replace β by $\beta + \lambda$ ('lambda') to obtain

$$\frac{1}{\omega}\sum_{boxes}(\Delta x)^3(\Delta v)^3\frac{e^{-\frac{1}{2}(\beta+\lambda)Mv^2}}{Z} = \frac{1}{Z\omega}V\left(\frac{2}{(\beta+\lambda)M}\right)^{3/2}K = 1 \quad (8.38)$$

for *any* positive number λ. Since this is true for any λ, it must also be true to first order in λ. Now, using (5.10):

$$e^{-\frac{1}{2}(\beta+\lambda)Mv^2} = e^{-\frac{1}{2}\beta Mv^2}e^{-\frac{1}{2}\lambda Mv^2}$$

$$= e^{-\frac{1}{2}\beta Mv^2}\left[1 - \frac{1}{2}\lambda Mv^2 + \cdots\right], \quad (8.39)$$

and, using the expansion (5.42) proved in solved problem (1) at the end of Chapter 5,

$$\left(\frac{1}{\beta+\lambda}\right)^{3/2} = \frac{1}{\beta^{3/2}}\left(1 + \frac{\lambda}{\beta}\right)^{-3/2}$$

$$= \frac{1}{\beta^{3/2}}\left[1 - \frac{3}{2}\frac{\lambda}{\beta} + \cdots\right]. \quad (8.40)$$

---

† $K = \pi^{3/2}$, with π, as usual, the ratio of the circumference to the diameter of a circle!

Finally, equating the terms linear in λ on the left and right hand sides of (8.38) we reach the wonderfully simple promised result

$$
\begin{aligned}
< E_{kinetic} > &= \frac{1}{Z\omega} \sum_{boxes} (\Delta x)^3 (\Delta v)^3 \frac{1}{2} M v^2 e^{-\frac{1}{2}\beta M v^2} \\
&= \left(\frac{3}{2\beta} \right) \left[\frac{V}{Z\omega} \left(\frac{2}{M\beta} \right)^{3/2} K \right] = \frac{3}{2\beta}.
\end{aligned}
\tag{8.41}
$$

In the last step, the normalization condition (8.35) has been used. You will verify that this is nothing more or less than (8.32).

The significance of this result was discussed at the end of the last section and the beginning of this one. We have proved that the reciprocal of the β in the Gibbs distribution is indeed equal to $k_B T$ where T is the gas temperature in kelvin. Gas temperature and absolute temperature are identical!

Effusion from a thermal enclosure

The Maxwell distribution contains much more information than just the average energy. To check that it does describe the real world, consider what it predicts about the molecules that emerge from a small hole in an enclosure containing a dilute gas and kept at constant

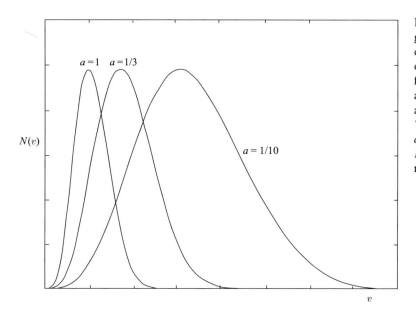

Fig. 8.3. Illustrative graphs of the number of molecules per unit of velocity effusing from an aperture in an oven at an absolute temperature T determined by $a = M/2k_B T$, where M is the mass of the molecules.

temperature. This process is called effusion. Apparatus can be constructed which measures the number of molecules – call it $N(v) \times \Delta v$ – that emerge from an aperture in a time Δt with speeds between v and $v + \Delta v$. If the hole is small enough, the energy carried away by the escaping molecules will be resupplied sufficiently quickly – by the device that is keeping the enclosure at constant temperature – to keep the system always in equilibrium. The observable quantity, $N(v)$ can then be calculated by summing the equilibrium distribution (8.33) over position and velocity 'boxes' which correspond to states of motion that carry a molecule out through the aperture in the prescribed interval of time. Without going through the details of this sum, we can argue that it contains a multiplicative factor of v^3. One factor of v comes from the fact that molecules a distance $v \Delta t$ from the hole can escape in a time Δt; the other two factors express the fact that the escaping particles can come from varying angles, i.e. that many velocities correspond to the same speed: the relevant velocity 'boxes' fill a segment of a spherical shell of average radius v and of thickness Δv. Since the surface area of a sphere is proportional to the *square* of its radius, this effect gives an extra factor of v^2. Thus the prediction is

$$N(v) \propto (av^2)^{\frac{3}{2}} e^{-av^2}, \tag{8.42}$$

where a, according to (8.33), is equal to $(M/2k_B T)$, and the symbol $\propto$ means, as before, 'is proportional to.' The constant of proportionality contains the area of the hole, the time Δt, and numerical constants that need not concern us. Figure 8.3 is a plot of (8.42) for various values of a. Note that smaller a means larger temperature, which has the intuitively plausible effect of generating 'effusing' particles with higher velocities.

The predicted behavior is obeyed by experiments within experimental error. The shape of the velocity distribution and the shift of the peak speed to higher values with increasing temperature are observed as predicted.

This is only one of many, many confrontations between theory and experiment which show that equilibrium is indeed described by the Gibbs distribution.

Solved problems

(1) <u>Entropy from partitioning requirements</u>. Generalize the requirement (8.8) to a system with I states with probabilities $p_1, p_2, \ldots p_I$ and hence deduce (8.7) in the same way as (8.9) was deduced from (8.8).

Solution: Since any number may be approximated as closely as desired by a fraction, we may assume without loss of generality that the probabilities are fractions, and put them all on the same (possibly large) denominator, L. In short, we may write $p_i = l_i/L$ for $i = 1, 2, \ldots, I$, where each l_i is an integer. Since the p_is must add up to unity, the l_is must add up to L. Now, as in the discussion preceding (8.8) we may think of the system as exploring L equally likely states of which groups of l_i are 'imaginary.' The generalization of (8.8) is

$$S(\frac{l_1}{L}, \frac{l_2}{L}, \ldots \frac{l_I}{L}) = \ln L - \sum_{i=1}^{I} \frac{l_i}{L} \ln l_i, \tag{8.43}$$

where we have subtracted from the entropy associated with L equally likely states ($\ln L$) the entropy associated with l_1 equally likely states multiplied by the probability (l_1/L) associated with this group of 'imaginary' states, and made the same subtraction for the groups of l_2, l_3, $\ldots l_I$ states. Multiply the $\ln L$ by unity written in the form $1 = (l_1/L)+(l_2/L)+\ldots(l_I/L)$, and use $\ln L - \ln l_i = -\ln(l_i/L)$ for each i to get

$$S(\frac{l_1}{L}, \frac{l_2}{L}, \ldots \frac{l_I}{L}) = -\sum_{i=1}^{I} \frac{l_i}{L} \ln \frac{l_i}{L} = -\sum_{i=1}^{I} p_i \ln p_i, \tag{8.44}$$

which is (8.7).

(2) <u>Inferring probabilities.</u> Consider an experiment in which the numbers one, two, and three are randomly drawn. If each number were equally likely, so that each had a probability $\frac{1}{3}$, the average number obtained observed after many trials would be

$$1 \times \frac{1}{3} + 2 \times \frac{1}{3} + 3 \times \frac{1}{3} = 2 \tag{8.45}$$

Suppose that after many trials this average is found instead to be 2.5. Use entropy considerations to assign probabilities to the three possible outcomes.

Solution: The method used in this chapter is sometimes touted as a general way of inferring probabilities from incomplete information, and called the 'strategy of least bias.' In Statistical Mechanics one has a microscopic picture in which thermal fluctuations are indeed trying out all possible assignments of probabilities, consistent with the constraint – of fixed total energy. If the experiment referred to in this problem is, say, the repeated rolling of a *unique* die with opposite pairs of faces marked 1, 2, and 3, such a picture is not available, and maximizing the entropy – which we shall now nonetheless do – has no precise meaning. A commonsense approach to probability, which I hope is one of the messages of this book, would suggest using results obtained in this way with considerable caution.

Call the probabilities we are looking for p_1, p_2, and p_3. Two conditions that must be obeyed by these quantities are

$$p_1 + p_2 + p_3 = 1, \tag{8.46}$$

and

$$1 \times p_1 + 2 \times p_2 + 3 \times p_3 = 2.5. \tag{8.47}$$

The first is the normalization condition (8.13) for $I = 3$. The second incorporates the given information about the observed mean, and is entirely analogous to (8.17). Two equations are not enough to determine the three unknown probabilities. For any particular experiment, the matter should rest here. However, if one can argue that one is dealing with an 'ensemble' of experiments all of which obey (8.46–47), then in analogy with the method of this chapter, maximizing the entropy – which in the present context is sometimes called the 'missing information function' – subject to the 'constraints' (8.46–47) produces the 'most probable' distribution. The relevant system of equations is then exactly (8.18–20) with $\epsilon_1 = 1$, $\epsilon_2 = 2$, and $\epsilon_3 = 3$, from which it follows using the same steps that led to (8.26) that the assignment of maximum entropy has the form

$$p_i = e^{\alpha} e^{-i\,\kappa}, \quad i = 1,\ 2,\ 3. \tag{8.48}$$

[We have used κ (Greek 'kappa') for a constant pure number, analogous to β in (8.26) but without its dimensions or physical meaning.] Equations (8.46 - 48) can now be solved† and give the answer

$$p_1 = \frac{1 + \sqrt{13}}{6(3 + \sqrt{13})} \approx 0.1162$$

$$p_2 = \frac{7 + \sqrt{13}}{6(3 + \sqrt{13})} \approx 0.2676 \tag{8.49}$$

$$p_3 = \frac{5 + 2\sqrt{13}}{3(3 + \sqrt{13})} \approx 0.6162.$$

You will verify that (8.46–47) are indeed obeyed by this set of numbers.

(3) <u>Ideal gas expansion at constant entropy.</u> Starting from the Gibbs distribution, show that the entropy S of N ideal gas molecules, each of mass M, occupying a volume V at a pressure p is given by the formula:

$$S = -N \ln \omega + N \ln V - \frac{3}{2} N \ln \frac{MN}{2\pi pV} + \frac{3}{2}N. \tag{8.50}$$

Here ω is the volume of the 'box' in position and velocity space needed to make sense of a dimensionless probability. Hence show that the expansion of an ideal gas at constant entropy is governed by the equation

$$p\,V^{\gamma} = \text{constant, where } \gamma = \frac{5}{3}. \tag{8.51}$$

Solution: For an ideal gas, the entropy per molecule is again obtained from the Gibbs distribution, except that now the sum over states has to done as on p. 139. From (8.33), we have

$$\ln p(\vec{x}, \vec{v}) = -\frac{1}{2}\beta M v^2 - \ln Z \tag{8.52}$$

---

† If you wish to check this for yourself, note that if $e^{-\kappa} = x$, then $e^{-2\kappa} = x^2$, and $e^{-3\kappa} = x^3$. Now, (8.48) and (8.46) imply $e^{\alpha} = (x + x^2 + x^3)^{-1}$, and (8.47) then shows that x is determined by solving the quadratic equation $x^2 - x - 3 = 0$. Since x like $e^{-\kappa}$ must be positive, $x = \frac{1}{2}(1 + \sqrt{13})$, and (8.49) follows.

From (8.35) – using [see the footnote on p.139] $K = (\pi)^{\frac{3}{2}}$ – we have

$$\ln Z = \ln V - \ln \omega - \frac{3}{2} \ln \frac{\beta M}{2\pi}. \tag{8.53}$$

Using (8.52) in the general expression for the entropy

$$S = -\frac{1}{\omega} \sum_{boxes} (\Delta x)^3 (\Delta v)^3 p(\vec{x}, \vec{v}) \ln p(\vec{x}, \vec{v})$$

$$= \ln Z + \beta \langle \frac{1}{2} M v^2 \rangle, \tag{8.54}$$

where the angular brackets mean an average over the Gibbs distribution. We know from (8.41) that the last term in (8.54) is $\frac{3}{2}$. The entropy for N molecules of an ideal gas is N times the entropy for one molecule. The only other substitution to get the quoted formula is the elimination of β in (8.54) via the ideal gas law $\beta = (N/pV)$.

Since $N \ln V = (3/2)N \ln(V)^{(2/3)}$, the expression (8.50) for the entropy may be written as

$$S = -N \ln \omega - \frac{3}{2} N \ln \frac{MN}{2\pi p V^{\frac{5}{3}}} + \frac{3}{2} N. \tag{8.55}$$

From this equation one sees at once that for a volume change at fixed entropy, it is necessary that the pressure also change in such a way that $pV^{\frac{5}{3}}$ remains constant. Note the difference with an expansion at constant temperature, which, according to the ideal gas law, requires pV constant. Using the ideal gas law in the last formula shows that for such an 'isentropic' expansion $TV^{2/3}$ is constant. A constant entropy expansion thus *cools* the gas. In simple physical terms, what is happening is that the mean kinetic energy per gas molecule – proportional via (8.41) to the temperature – is decreasing because energy is being used to move the boundaries of the system. This use of energy is called 'work,' and will be fully explained in the next chapter.

(4) <u>Temperature dependent entropy of a two-level system.</u> Consider a thermodynamic system composed of a large number of weakly interacting subsystems each of which has two states of motion with energy 0 and ϵ. Show that the entropy per subsystem in equilibrium, at temperature $T = (k_B \beta)^{-1}$, can be put in the form:

$$S = \frac{\ln(1 + e^{-\beta\epsilon})}{1 + e^{-\beta\epsilon}} + \frac{\ln(1 + e^{\beta\epsilon})}{1 + e^{\beta\epsilon}}. \tag{8.56}$$

Plot S as a function of $(1/\beta\epsilon) = k_B T/\epsilon$. Values are given in Table 8.1. Note that you can work out, and in fact already know, the values of S for $T = 0$, and for $k_B T \gg \epsilon$. Also note that $\ln 2 \approx 0.69$.

Say in words what physical sense you can make of the curve you have drawn.

Solution: Let us label the states of motion 1, energy $\epsilon_1 = 0$, and 2, energy $\epsilon_2 = \epsilon$. Then the Gibbs distribution for the probabilities of occupation at absolute temperature T is an especially simple case of (8.28), namely

$$p_1 = \frac{1}{1 + e^{-\beta\epsilon}}, \qquad p_2 = \frac{e^{-\beta\epsilon}}{1 + e^{-\beta\epsilon}}. \tag{8.57}$$

The second term can by rewritten by using $e^{-x} = (1/e^x)$. Then, substituting (8.57) into the general expression for the entropy

$$S = -p_1 \ln p_1 - p_2 \ln p_2 \tag{8.5}$$

gives the desired expression.†

S has been plotted according to the given table of values in Fig. 8.4. Note that at $T = 0$, when $p_1 = 1$ and $p_2 = 0$, the entropy, corresponding to complete order, is 0, and that as $T \to \infty$, when both p_1 and p_2 approach $\frac{1}{2}$, S approaches the value for maximum mixing between two states, i.e. $\ln 2$. This is as you expect: heating the system disorders it.

Table 8.1. *Values for problem 4.*

$k_B T/\epsilon$	$S(T)$	$k_B T/\epsilon$	$S(T)$	$k_B T/\epsilon$	$S(T)$	$k_B T/\epsilon$	$S(T)$
0.125	0.003	0.250	0.09	0.375	0.24	0.500	0.37
0.75	0.51	1.00	0.58	1.25	0.62	1.50	0.64
1.75	0.65	2.00	0.66	2.50	0.67	3.00	0.68

Fig. 8.4. Temperature dependence of the entropy of a two-state system.

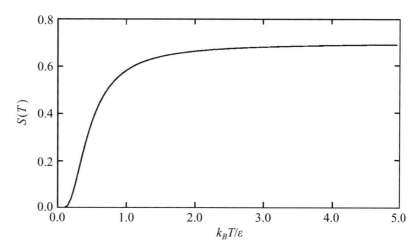

†Some may notice an apparent inconsistency here. Below (8.17), it is remarked that at least three independent probabilities are needed to derive the Gibbs distribution. Yet, we are now blithely applying it to a subsystem with only two levels. The paradox is resolved by considering two such systems taken together as the basic unit. There are now four states with probabilities P_1, P_2, , P_3, P_4, and energies 0, ϵ, ϵ, and 2ϵ, corresponding to both systems de-excited, one or the other excited, and both excited. The Gibbs distribution can unquestionably be applied to this four-level subsystem. A little algebra then shows that in equilibrium $P_1 = (p_1)^2$, $P_2 = P_3 = p_1 p_2$, and $P_4 = (p_2)^2$, where p_1 and p_2 are given by (8.57). Because the probabilities for the composite are products, we see that it is consistent to assign independent probabilities to each two-level system.

The meaning of the curve is as follows. As the temperature increases, the entropy shows a smooth transition from complete order to the maximum possible disorder. The cross-over occurs at a temperature roughly determined by the equality $k_B T = \epsilon$, which – as (8.57) shows – separates regions of thermal equilibrium corresponding to very dissimilar and almost equal probabilities.

In the next chapter it will be shown that entropy differences are related to inputs of heat. As a consequence, entropy differences are measurable. In fact, a temperature-dependent entropy very similar to that just calculated is observed in experiments on certain magnetic crystals – such as Cerium Magnesium Nitrate. The model of almost independent two-state subsystems applies to these materials because they contain well separated 'electron spins,' associated with the Cerium atoms, which can take on two values, parallel and antiparallel to an applied magnetic field.

9

Heat, work, and putting heat to work

A man hath sapiences thre
Memorie, engin and intellect also
Chaucer

Engines are, etymologically and in fact, ingenious things. By the end of this chapter we shall understand heat-engines, which are devices that convert 'heat' – an intuitive idea to be made precise here – into pushes or pulls or turns. First, however, we need to systematize some of the things we have already learnt about energy and entropy, and thereby extract the science of Thermodynamics from the statistical viewpoint of the last chapter.

Thermodynamics is a theory based on two 'Laws,' which are not basic laws of nature in the sense mentioned in Chapter 1, but commonsense rules concerning the flow of energy between macroscopic systems. From the point of view we are taking here, these axioms follow from an understanding of the random motion of the microscopic constituents of matter. It is a tribute to the genius of the scientists – particularly Carnot, Clausius, and Kelvin – who formulated thermodynamics as a macroscopic science, that their scheme is in no way undermined by the microscopic view, which does, however, offer a broader perspective and open up avenues for more detailed calculations. (For example, effusion – discussed at the end of the last chapter – lies beyond the scope of thermodynamics.)

Work, heat, and the First Law of Thermodynamics

The First Law of Thermodynamics is about the conservation of energy. The concept is the one we encountered in Chapter 6 where, you will recall, we found that a falling body could be thought of as converting potential energy into kinetic energy – the sum of the two remaining constant. The important distinction for thermodynamics is somewhat different. It is between the energy, both kinetic and potential, associated with random molecular agitation, and the energy, also both kinetic and potential, of a collective motion of a large number of molecules, such as a moving piston or lever.

The technical word for changes of energy associated with imparting motion to macroscopic objects is *work*. The simplest example of such a process is the expansion of a gas against a piston, illustrated schematically in Figure 9.1. Let the pressure of the gas be p and the area of the piston A, so that the force, F, perpendicular to the face of the piston is pA. Then, if the piston moves a infinitesimal amount Δx in the direction of the force the product of these quantities is *defined* to be the element of work, call it ΔW, done on the piston. Since $A\Delta x$ is the change in the volume, V, of the gas, we can also say

$$\Delta W = F\Delta x = p\ A\Delta x = p\Delta V. \tag{9.1}$$

If the change in volume is small the pressure will be constant, and the work done on the piston is the pressure multiplied by the increase in volume of the gas. If the force of gas pressure is the only one on the piston, the work (9.1) is converted, via Newton's second law, into an increase of kinetic energy of the piston. (The argument is identical to the one given at the end of Chapter 6, where the increase of kinetic energy of the falling apple could have been viewed as due to the work done on the apple by the gravitational force.) We can also imagine the piston connected to a cord carrying a weight, in which case the work done by the gas could be converted into potential energy of the weight. To keep things as simple as possible, we shall neglect friction acting on the piston or the seal to the cylinder in the balance of energy, as something which could if necessary be put in later.

In conventional thermodynamics one now invokes conservation of energy to argue that an increase in energy of the piston can only come from a decrease in energy of the gas, which argument is strictly

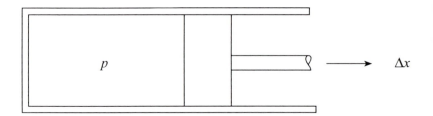

Fig. 9.1. Movement of a piston via the expansion of a gas.

correct. A thermodynamic system – by which one means a macroscopic quantity of matter contained within a boundary – whose energy is changed *only* by work done on other macroscopic bodies thus obeys the equation

$$\Delta E = -\Delta W = -p\Delta V, \tag{9.2}$$

where the last equality only applies to a gas or liquid exerting a uniform pressure, p, on its containing walls.

Since we have a microscopic picture at our disposal, it is instructive to think about how such an energy transfer occurs. Here the molecular model of an ideal gas used in the discussion of gas pressure in Chapter 7 is helpful. It is illustrated in Fig. 9.2, in a manner intended to be reminiscent of Fig. 7.1. The smooth 'wall' on the right is now imagined to be moving with a very small constant velocity u. A colliding molecule will preserve its component of velocity parallel to the wall but its perpendicular speed will be decreased by $2u$.

Why? Consider the collision from the point of view of someone moving with the wall. This 'observer' is moving at constant velocity and her perspective is just as good as the one illustrated as far as the application of Newton's Laws is concerned: accelerations, which are all that Newton cares about, are identical from both points of view. As viewed from the wall, the molecule approaches with a perpendicular velocity $v_\perp - u$, which, as a result of the collision and exactly as in Fig, 7.1 is exactly reversed. To go from the 'frame of reference' of the wall to the one illustrated in the figure one must add u to the perpendicular component of the velocity of every particle moving towards the wall

Fig. 9.2. Collision of a molecule with a moving wall

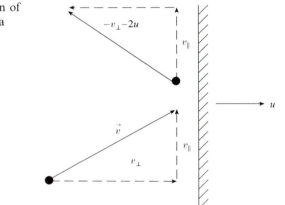

and subtract the corresponding amount for particles moving away from the wall.† This leads to the labeling shown in the figure.

Let us calculate the loss of kinetic energy of the colliding molecule. Using the Pythagorean theorem to calculate the square of the speed of the molecule before and after the collision, we have, to linear order in u:

$$\Delta\epsilon_{per\ collision} = \frac{1}{2}M[v_\parallel^2+(v_\perp-2u)^2]-\frac{1}{2}M[v_\parallel^2+v_\perp^2] \approx -2Mv_\perp u. \quad (9.3)$$

Equation (9.3) will allow us to calculate the connection between the energy lost by the slowly expanding gas and its pressure in two instructive ways. First, we shall use the kinetic method of Chapter 7. Consider the energy lost by particles moving towards the wall with perpendicular velocity $v_\perp$ while the wall moves a distance $\Delta x = u\ \Delta t$. This energy loss is given by the product of factors $(-2Mv_\perp u) \times (\Delta t/\tau(v_\perp))$. The first factor comes from (9.3); the second is the mean number of collisions in the time Δt, given that $\tau(v_\perp)$ is the mean time between collisions, as defined in (7.6). [This and the next several paragraphs make precise use of numbered formulas in chapters 7 and 8. The reader should please refer back to these equations.] Note that $u\Delta t$ (the distance Δx the wall moves) comes out as a factor from this product, and that what is left is minus the force given in (7.5). Summing over all the molecules, exactly as in (7.6–7.8), translates what we have just shown into the statement that the kinetic energy *lost* by the gas when its volume increases the small amount ΔV is $p\Delta V$, where p is the average gas pressure given by the kinetic theory of Chapter 7 (7.8), and $\Delta V = A\Delta x$, with A the area of the piston. Thus, the work (9.1) done on the piston is indeed equal to the loss of total energy of the gas.

Now, the energy being transferred from the gas to the piston in the form of work is in a quite fundamental sense different from the energy exchanged between one subsystem and another in (8.19). In that equation we imagined changing the energy of a subsystem by changing the probabilities of occupation of the states of motion. Here, as (9.3) begins to indicate, we are changing the energy of a state of motion. This can be made clearer by calculating the reduction in

---

† The words 'observer' and 'frame of reference' are most commonly heard in connection with Einstein's Special Theory of Relativity in which the speed of light is an absolute constant. [If Einstein had named his theory the Theory of the Absolute Constancy of the Speed Light, a lot of metaphysical confusion associated with the word 'relativity' might have been avoided.] That theory, which applies when speeds near that of light are involved, has an unobvious and most interesting transformation between different frames of reference. It is not of relevance to the problem at hand, where the argument given, based, if you wish, on the Newtonian Relativity of Newton's Equations, is a useful shortcut.

kinetic energy of a molecule when the gas contained by the piston is expanded *very slowly* by a volume ΔV.† In Fig, 9.2, suppose that the length of the cylinder from the stationary back wall to the piston is L. Imagine, as is utterly reasonable, that the speed of motion of the piston, u, is very very much smaller than the velocity of the molecule perpendicular to the piston face, $v_\perp$. In the time Δt that it takes for the piston to move a small amount, the molecule will have made many traversals of the cylinder. In fact the number of collisions with the piston, N, will be the time Δt divided by the time, $2L/v_\perp$, it takes for one traversal back and forth. From (9.3), the loss of kinetic energy of the molecule is thus

$$\Delta\epsilon = \Delta\epsilon_{per\ collision} N = -2Mv_\perp u\frac{v_\perp}{2L}\Delta t. \tag{9.4}$$

Now $u\Delta t$ is, as before, the distance Δx moved by the piston. Let us call $\frac{1}{2}Mv_\perp^2$, the kinetic energy associated with motion perpendicular to the container wall, $\epsilon_\perp$. Then (9.4) says

$$\Delta\epsilon = -2\frac{\epsilon_\perp}{L}\Delta x = -2\frac{\epsilon_\perp}{V}\Delta V, \tag{9.5}$$

where V and ΔV, the volume and change of volume respectively, are as above LA and $\Delta x\, A$, with A the area of the piston.

It is just a short step from (9.5) to a proof of the equality between the loss of energy of the gas and the work done on the piston using the statistical method of Chapter 8. Consider (8.17) for the average energy, E, per subsystem – which in the present case is a molecule. For the very slow process we are considering, the energy of *every* colliding molecule is changed as in (9.5) so that even though the energies of the states are changing, the number of molecules corresponding to any state – and thus the probability to be associated with that state – is not changing. The change in average energy per molecule can thus be expressed as

$$\Delta E = \sum_i (\Delta\epsilon_i)p_i = -\frac{2\Delta V}{V}\sum_i (\epsilon_i)_\perp p_i, \tag{9.6}$$

where we have used (9.5) to get the last form.

In equilibrium, the p_is are given by the Maxwell distribution (8.33) in which speeds in all directions are equally likely. Thus the average of the kinetic energy perpendicular to the piston, which occurs in (9.6), is one-third of the average kinetic energy. Since kinetic energy is all

---

† At this point the statistical discussion that follows becomes less general than the kinetic argument of the last paragraph where the kinetic pressure on the wall did not necessarily have to correspond to equilibrium. Statistical mechanics as we have developed it, which we shall now use to obtain (9.7) when p is the gas pressure in equilibrium, only deals with equilibrium and small deviations from it.

that matters in an ideal gas, this is one-third of the average energy. Using these facts we may write

$$\left(\frac{\Delta E}{\Delta V}\right)_{fixed\ p_i s} = -\frac{2E}{3V} = -\frac{T}{V} = -p. \tag{9.7}$$

To obtain the third expression we have used (8.32) for the average kinetic energy per molecule, and in the last form the ideal gas law scaled down to account for the fact that we have been considering the contribution of a single molecule.† It is worth mentioning that the fixed p_is indicated in (9.7) imply by via the definition (8.7) that the *entropy* is fixed during the change in volume we have contemplated.

In general, then, when a thermodynamic system does work on a macroscopic external object, the energy of the system is reduced by the amount of that work. Conversely, the energy of a thermodynamic system is increased by the amount of work done on it by an external agency – just imagine reversing the direction of the piston in our example.

It is not only by doing work that a system can lower its energy. A hot gas in a container with rigid walls that are not perfectly insulating will gradually acquire the lower temperature of its cooler surroundings, transferring energy to them in the same way as the equilibrating subsystems in the discussion in the last chapter transferred energy to each other. This form of transferred energy is called heat. Conservation of energy then requires that the heat input into a system be the increase in energy plus the work done by the system. This statement of conservation of energy is called The First Law of Thermodynamics. It is expressed in the formula

$$\Delta E = \Delta Q - \Delta W. \tag{9.8}$$

Here the left hand side is the *increase* in the energy of the system, ΔQ is the *inflow* of heat, and ΔW is the work done *by* the system on its surroundings. The italicized words explain how the accounting must be done in order for the signs in (9.8) to be correct: work done by the system *reduces* its energy. (Note that each of the quantities in (9.8) can be positive or negative – a decrease of energy is a negative increase, etc.)

Since the first law is merely a statement of conservation of energy, it must in the special case of small changes from equilibrium be

† The absolute temperature T in equation (9.7) and from now on is in the energy units advocated in Chapter 7, so that the energy previously written $k_B T$ is now simply T. The ideal gas law (7.12) is $pV = NT$ in these units, or $p = (T/V)$ per molecule.

contained within the statistical expression for the energy of a thermodynamic subsystem, i.e., within (8.17). That expression depends on the energies, and on the occupation probabilities of the possible states of motion; it will change when either of these two quantities changes. To leading order in small changes, (8.17) yields

$$\Delta E = \sum_{i=1}^{I} (\epsilon_i \, \Delta p_i + \Delta \epsilon_i \, p_i). \tag{9.9}$$

The second term, corresponding to a change of the energies of the states of motion, has already been identified with work – minus the work done by the system to be precise – and thus corresponds to the second term in (9.8). The first term, a change of energy associated with a redistribution of occupation probabilities, is exactly how we allowed subsystems to exchange energy in our discussion of equilibrium. This, then, is the microscopic interpretation of what in thermodynamics is called an input of heat. Thus for small deviations from equilibrium we may write for the input of heat

$$\Delta Q = \sum_{i} \epsilon_i \Delta p_i. \tag{9.10}$$

To get the first law from (9.9) one has to add expressions of this form for each of the subsystems making up a thermodynamic system.

Note that (9.9) has given considerable insight into (9.8). In thermodynamics, once one accepts that the average energy of a subsystem is determined by its state of equilibrium, (9.8) merely defines heat flow. In (9.9) and (9.10), however, heat input is instructively seen to have the effect of repopulating the energy states, whereas when work is done it is the energy states themselves that are changed without transitions of subsystems from one state to another. It is worth remarking that, as (9.9) makes clear, heat and work are concepts that occur only as differences. How much heat or work a system contains is not a sensible question. In contrast it does makes sense to ask what the average energy of a system is – up to an overall constant related to the reference point from which one choses to measure potential energy.

Heat, entropy, and the Second Law of Thermodynamics

That ice in a glass of water melts and cools the drink may seem too obvious to merit thought. And yet, nothing in the first law prevents the opposite from happening. Take a motion picture of the process and run it backwards. The mist on the glass disappears, the glass and liquid warm, the ice cubes get bigger. Energy is conserved at every step, but our experience tells us that what we are witnessing would not go of itself. This is the domain of the Second law. It gives a systematic answer to the question of what processes involving the flow of heat and work happen spontaneously. Whereas the first law is contained in Newton's equations in a perfectly straightforward way, the second law applies only to thermodynamic systems, i.e., to systems with many, many sub-parts. There are various ways of phrasing the second law, but you are in a position to cut through to the heart of the matter. Following the discussion in Chapter 8, one would say that spontaneous processes tend towards a new equilibrium, and that equilibrium means the maximal exploration of the available states of motion – i.e. the state of maximum entropy. Indeed, the most concise statement of the second law is the following. The total entropy of an *isolated* macroscopic system cannot decrease. In symbols, taking S to be the entropy, the second law reads

$$\Delta S \big|_{isolated\ system} \geq 0. \tag{9.11}$$

When the entropy change is zero, one speaks of a reversible process. From the probabilistic point of view, this limit describes changes which do not decrease order, for example those which preserve the overall equilibrium of an isolated system.

In the remainder of this section we shall show that (9.11) implies that without outside intervention heat can only flow from a hot object to a cold object and not in the opposite direction. Later in the chapter, we shall also find that (9.11) has the extremely interesting consequence of placing limitations on the conversion of the energy of chaotic molecular motion into the energy of directed large-scale movement, which conversion is, of course, what heat engines do.

Much can be learnt by studying small departures from equilibrium. In (9.9) we considered small changes of energy. Now let us contemplate small changes in entropy near equilibrium. Start by evaluating (8.16) for small deviations from equilibrium, in which case the p_i^0s are given by the Gibbs distribution (8.28), from which

$$\ln p_i^0 = -\beta \epsilon_i - \ln Z. \tag{9.12}$$

Substituting this into (8.16), we find

$$\Delta S = \beta \sum_{i=1}^{I} \epsilon_i \Delta p_i. \tag{9.13}$$

[What has happened to the $\ln Z$? Since it is independent of i, it comes out as a factor multiplying the sum $\sum \Delta p_i$ which has numerical value zero – as was discussed in general below (8.16).]

Very interestingly, the right hand side of (9.13) is proportional to the first term in (9.9), interpreted as an input of heat via (9.10). Thus for small deviations from equilibrium we have †

$$\Delta S = \beta \Delta Q = \frac{\Delta Q}{T}. \tag{9.14}$$

Simple though it has been for us to derive, this equation is one of the most fundamental in thermodynamics. [Its derivation using macroscopic reasoning is an intellectual tour de force which, as I mentioned in the last chapter, has bedevilled generations of science and engineering students.] It is worth re-emphasizing that it applies to *small* deviations from *equilibrium*. Although we have obtained it for a subsystem, it also applies to a collection of weakly interacting subsystems, because entropy and energy are then both additive quantities.

We can now use (9.11) and (9.14) to show that the identification of spontaneous processes with increasing entropy makes intuitive good sense. Suppose that two systems with different temperatures are put into contact with one another, isolated from the rest of the world, and allowed to exchange a small amount of heat. Does the entropy of the composite increase or decrease? Here one has to watch signs. Using (9.14) one has

$$\Delta S = \Delta S_A + \Delta S_B = \Delta Q_{B \to A}[\beta_A - \beta_B]. \tag{9.15}$$

Now, $\Delta Q_{B \to A}$ is the heat *input* into system A. Let's take this to be positive. Then, according to (9.15), the total entropy increases if β_A is greater than β_B. An increase of entropy means more disorder, or more mixing, i.e., a tendency towards a new equilibrium. This requires, using the reciprocal relationship between β and temperature, that the temperature of system B be higher than the temperature of system

† For purely historical reasons the entropy is usually defined with a multiplicative factor of k_B, so that it then has units of energy per kelvin, and the last though not the second form of (9.14) becomes correct both with my conventions (in which entropy is a pure number with no dimensions and temperature like heat has the units of energy) and in the more common – but less rational – form, with the absolute temperature then in K.

A when heat flows into *A*. [Please run through this to make sure I haven't cheated.] But this makes a lot of sense. The tendency towards a new equilibrium is, as expected, associated with the flow of heat from the hotter (for which read higher temperature) to the colder (lower temperature) system.

Thermodynamic reasoning: an example

Equations (9.8) and (9.11) are complete statements of the first and second laws of thermodynamics. In thermodynamics *per se* the energy and entropy of a system are not calculated from a microscopic model but are posited to be determined by the macroscopic parameters which determine the equilibrium, the volume and the temperature for example. Differences in energy and entropy are then calculated using (9.7) and (9.14). Using these relations in (9.8) then gives for small deviations from equilibrium in a gas†

$$\Delta E = T \Delta S - p \Delta V. \tag{9.16}$$

It is instructive to use (9.16) to calculate the increase in entropy in a volume doubling expansion of an isolated ideal gas. This is the irreversible process discussed on p. 127 where we argued that the doubling of the volume of an ideal gas increases its entropy by $N \ln 2$, N being the number of molecules. If this were to happen by the opening of a valve connecting a container of gas to another evacuated one, the entropy would be *irretrievably* increased by this amount.‡ Thus $N \ln 2$ is the difference in entropy of two equilibrium states, the latter, however, being reached after an irreversible expansion.

How in the world is anything about such a process to be calculated from (9.16), which only applies to small deviations from equilibrium? The answer is that the transition between the initial and final equilibrium states could also be carried out reversibly, i.e., through a sequence of intermediate equilibrium states. It is very instructive to think through how to do that. Evidently, a slow expansion of the gas in thermal isolation against a piston will not reach the desired condition at the end, because at every step of the way work is done against the piston, and the energy of the gas is reduced while the entropy remains

---

† Although the kinetic and statistical derivations of the expression $p \Delta V$ for the work done by a gas neglected interactions between molecules and were thus restricted to ideal gases, the argument based on conservation of energy shows that it applies to gases in general.

‡ Note the implicit assumption on p. 127 that the velocity distribution is the same before and after the expansion, which assumption we can now justify on grounds that the temperature and thus the Maxwell distribution (8.28) is the same in the initial and final equilibrium.

constant. Because of this reduction of energy, and thus through (8.32) temperature, a volume doubling expansion done slowly in thermal isolation leads to a different (colder) final equilibrium state than that reached after a free expansion also done in thermal isolation. If we want a reversible process whose end result is the same as that after free expansion, we have to replace, by the injection of heat, the energy that has been transferred out of the system to slowly moving walls. One could, for example, do this at every step of the process: making a small expansion, adding enough heat to bring the energy of the gas back up to its original value, and so on. Equation (9.16) tells how the two step process works. In the first small step when there is no flow of heat ΔS is zero, and $\Delta E = -p\Delta V$. In the second step, the input of heat $T\Delta S$ is to exactly compensate for the loss of energy in the first step. Thus

$$\Delta S = -\frac{\Delta E}{T} = \frac{p}{T}\Delta V = \frac{N}{V}\Delta V = N\frac{\Delta V}{V}. \tag{9.17}$$

The first two equalities apply to any gas subjected to these two compensating processes. In going from the third to the fourth form, we have used the ideal gas law $pV = NT$, which we derived in Chapter 7 using the kinetic theory, but which from the point of view of thermodynamics is an empirical law, obtained for example from experiments using a gas thermometer.

When ΔV is very much less than V, (5.17) shows that the last form of (9.17) is equivalent to

$$\Delta S = N\ln(1 + \frac{\Delta V}{V}) = N\ln(\frac{V + \Delta V}{V}). \tag{9.18}$$

Proceeding step by step, and recalling from (5.13) how to add logarithms,† one sees that the total entropy change after a very large number of such tiny two-step processes is given by

$$S_f - S_i = N\ln(\frac{V_f}{V_i}). \tag{9.19}$$

where S_f, V_f and S_i, V_i are the final and initial values. A volume doubling expansion at constant energy of an ideal gas has thus been shown by thermodynamic reasoning to increase its entropy by $N\ln 2$, where N is the number of molecules in the gas.

As a last remark on this instructive example, note that an entropy

---

† For example, after two small expansions from V to $V + \Delta V$ and then from $V + \Delta V$ to $V + 2\Delta V$ the total entropy change divided by N is given by

$$\ln(\frac{V + \Delta V}{V}) + \ln(\frac{V + 2\Delta V}{V + \Delta V}) = \ln\left[\frac{V + \Delta V}{V}\frac{V + 2\Delta V}{V + \Delta V}\right] = \ln\frac{V + 2\Delta V}{V}.$$

increase in the gas has here been achieved reversibly. Where did this entropy come from? To answer this question we have to inquire about the origin of the many heat inputs all at the temperature T. Suppose that these are provided by a source itself always in equilibrium with absolute temperature T, and so large that its temperature remains constant while the total amount of heat $T(S_f - S_i)$ is extracted reversibly from it.† Then, no entropy is irretrievably generated in either the gas or the heat source, but $N \ln 2$ of entropy flows from the reservoir to the gas.

Heat engines

In the last two chapters we talked about equilibrium, a quiescent condition in the large which might seem to have nothing whatsoever to do with the rushing gases, valves, and exhausts of machines such as steam engines. And yet, the understanding we have developed of equilibrium and small deviations from equilibrium will help crucially in the task of this section. Even though this understanding gives no insight into the detailed workings of any engine, it will allow us to follow, in complete detail and considerable depth, a simple but seminal analysis of an ideal engine against which all real ones can be compared. From a conceptual point of view this ideal engine – the simplest that has been devised – goes back to very beginnings of thermodynamics. Called a Carnot engine, it is based on a cyclic process that takes heat from a high temperature 'source,' converts some of that heat into work, and rejects the remaining heat into a low temperature 'sink.' Carnot invented his engine to answer the question: How efficiently can heat be converted into work? We are now ready to reproduce his brilliant analysis concisely.

A key word above is 'cyclic.' A Carnot engine has a working substance, like the steam in a steam engine, which absorbs and gives up the heat, and does the work. In the course of an operational cycle, the working substance is returned to the thermodynamic state it started from. In analyzing the engine, one must consider complete cycles, whose net effect is the transfer of heat and the conversion of some of it into work. Otherwise the borrowing and repayment of energy by the working substance will obscure the basic energy balance.

In the original analysis of the Carnot cycle, an ideal gas was used as the working substance. As a matter of fact, entropy, as a slightly

---

† Such sources of heat are called a thermal reservoirs. They will occur frequently in the next section.

mysterious macroscopic concept, emerged from an extension by Clausius of Carnot's analysis. We shall now see that, once one understands entropy, the working substance in a Carnot engine becomes irrelevant. As a matter of fact, the analysis becomes straightforward.

Consider the situation schematized in Fig. 9.3. In one cycle, let Q_1 be the heat taken from the high temperature heat source at absolute temperature T_1, Q_2 the heat given up to the low temperature heat sink at absolute temperature T_2, and W the work done.

Now the first law, i.e. conservation of energy, requires:

$$W = Q_1 - Q_2. \tag{9.20}$$

This hardly needs comment. It is (9.8) summed up over one cycle for the working substance, which has the same energy at the end and start of each cycle. Please verify that the sign on the right is consistent with the manner in which heat inputs and outputs have been here defined.

Let us consider the *closed system* consisting of the working substance *and* the two heat reservoirs, and compare its entropy at the start and end of one cycle. By hypothesis the reservoirs are sufficiently large that their temperatures are negligibly changed by the flows of heat Q_1 and Q_2. Thus, (9.14), which in general is restricted to small changes from equilibrium, is applicable to the reservoirs for these finite quantities of heat. Also, since the working substance will have been returned to its original state, its net change of entropy is zero. According to the second law (9.11), the increase in the entropy of the low temperature

Fig. 9.3. Diagram illustrating a Carnot engine. The two rectangles represent heat reservoirs at the absolute temperatures indicated. The circle represents the working substance which in each cycle absorbs and emits heat and does work, as illustrated, and discussed in the text.

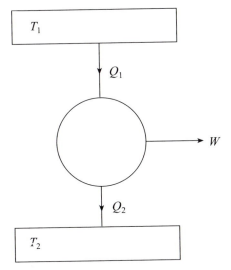

reservoir, Q_2/T_2, minus the decrease in entropy of the high temperature reservoir, Q_1/T_1, cannot be less than zero. In symbols we therefore have

$$\Delta S = \frac{Q_2}{T_2} - \frac{Q_1}{T_1} \geq 0 \qquad (9.21)$$

Equations (9.20) and (9.21) are all there is to the analysis of the Carnot cycle. The *efficiency*, η (Greek 'eta'), of an engine is defined in a fairly natural way to be the work done per cycle divided by the heat input per cycle. Solving (9.21) and (9.20), one finds

$$\eta = \frac{W}{Q_1} = \frac{Q_1 - Q_2}{Q_1} = 1 - \frac{Q_2}{Q_1} \leq 1 - \frac{T_2}{T_1} \qquad (9.22)$$

This is very interesting. We are learning that an engine operating between two temperature reservoirs can only convert a certain fraction of its heat input into work. In fact, since reversible processes [with $\Delta S = 0$] correspond to transitions between equilibrium states, they must be conducted very slowly. To run at any speed, a real engine must have some irreversibility, and thus an efficiency less than the ideal limit.

The inequality (9.21) teaches us something that is by no means intuitively obvious. It tells us that even if we oversimplify matters and ignore friction and all other sources of thermodynamic irreversibility the efficiency of an engine working between two heat reservoirs cannot exceed the expression given on the right of (9.22). Typically the heat input to an engine comes from burning some chemical fuel or from capturing the sun's rays. The first law in no way forbids the conversion of all the heat into useful work. The second law makes such a neat trick impossible, which is a pity because otherwise we could happily harness the immense amount of thermal energy in a large body of water at one temperature to run machines for us.

There is one, quite artificial, situation in which a Carnot cycle converts its entire input of heat into work. From (9.22) we see that this occurs if the heat sink is at the absolute zero of temperature, i.e. $T_2 = 0$. No such heat sink is available, even in outer space it turns out. As an abstract thought, however, the meaning of (9.22) when $T_2 = 0$ is fairly obvious. In order to extract *all* the heat energy of a working substance in the form of useful work, the substance must be completely de-excited, which is precisely the meaning of cooling it to the absolute zero of temperature.

Apart from this unreal limiting case, we can summarize what we have deduced from the principle of no spontaneous decrease of entropy as follows. It is impossible to devise a process working in a cycle that has no other effect than the complete conversion of heat from a single temperature reservoir into work.

Taken as an axiom about macroscopic heat flows, the last sentence is Kelvin's statement of the second law. It can be used, together with conservation of energy (the first law), to construct an entirely large scale, brilliantly logical basis for thermodynamics, without, however, the intuitive insight into entropy provided by the molecular viewpoint.

Refrigerators and heat pumps

A heat pump is a device that moves heat from a low temperature reservoir to a high temperature reservoir. A household refrigerator is the most common example of such a machine. Your intuition tells you that the transfer cannot occur spontaneously, and your intellect is now able to refer to the calculation that led to (9.15) to show that in an isolated system such a spontaneous transfer, corresponding to a macroscopic decrease of entropy, does not happen. Thus, an input of work is needed.

One immediately asks a question analogous to the one about engines in the last section. What is the most efficient refrigerator imaginable? In other words, what is the minimum amount of work that must be done to move a given amount of heat out of a low temperature reservoir? Again, one has to think about complete cycles of some device, in order not to become confused with temporary borrowings and repayments of energy. Again, the answer is provided by Carnot's cycle, this time, however, run in such a way that all the arrows in Fig. 9.3 are reversed, or, equivalently, that Q_1, Q_2, and W, as shown, are all negative. Let us define the 'coefficient of performance,' call it E, of the cycle to be the heat pumped *from* the low temperature reservoir per cycle divided by the work done *on* the heat pump per cycle. I invite you to show, from the requirement that the total entropy of the reservoirs cannot decrease at the end of one cycle, that

$$E = \frac{Q_2}{W} \leq \frac{T_2}{T_1 - T_2} \tag{9.23}$$

The conventions are those of the figure: T_2 is the low temperature and T_1 the high temperature. It is clear from the formula that a heat pump is most successful when the temperature difference through which the heat is to be moved is small.

Another use of a heat pump, perhaps not as familiar as a household refrigerator, is as a device for heating a house, in which case the low temperature reservoir is the outdoors, and the high temperature reservoir is the inside of a house. A little thought will convince you that in principle this is a more efficient way of heating than by irreversibly converting electrical or chemical energy to heat in the interior of the house. Equation (9.23) and the remark following it show that this use of a heat pump is a better idea in mid-winter Arizona than in mid-winter Siberia.

Evidently, the coefficient of performance (9.23) must be finite, even for a reversible cycle. Thus we have used the principle of no spontaneous decrease of entropy to prove the following statement. No process is possible which has no other effect than the transfer of heat from a low temperature to a high temperature reservoir within an isolated system. This is called Clausius' statement of the second law. It is equivalent to Kelvin's statement, discussed above.

This is as far as I shall take the subject of engines and refrigerators. The cycles that are used to make practical machines are more complicated than the Carnot cycle, and often involve more than two temperatures. What you have learned, however, in the matter of converting heat into work efficiently, or of moving heat from low to high temperatures efficiently, is to minimize irreversibility, i.e. the creation of entropy. The educated conservationist's slogan is not Conserve Energy, which is unavoidable, but the perhaps less catchy but correct Don't Generate Entropy!

Solved problems

(1) <u>Carnot's Cycle</u>. In the text we have discussed the Carnot cycle in a very general way, and obtained the correct expression for its efficiency. It is instructive, nonetheless, to consider the original Carnot engine, in which the working substance is an ideal gas. The cycle consists of the following four steps:

    (A) The ideal gas, initially at pressure p_1, volume V_1, and temperature T_1 is slowly expanded at constant temperature – 'isothermally' – to pressure p_2 and volume V_2.

    (B) It is then slowly expanded in thermal isolation – 'adiabatically' – to pressure p_3, volume V_3 and temperature T_2

    (C) Next, it is compressed isothermally to pressure p_4 and volume V_4.

    (D) Finally, it is compressed adiabatically to pressure p_1, volume V_1 and temperature T_1. It has now returned to its original state, ready for the start of its next cycle.

    Note that steps (A) and (C) occur at constant temperatures T_1 and

T_2 respectively, while steps (B) and (D), occur at constant entropies S_1 and S_2, because thermal isolation means no heat flow and thus via (9.14) for a slow process no change in entropy. For an isothermal expansion, the ideal gas law $pV = NT$ requires that the product pV stay constant; for an adiabatic (constant entropy) expansion (8.55) applies, so that $pV^{5/3}$ is constant. It will be also useful to note from (8.55) that the entropy of the gas may be expressed as

$$S = \text{const.} + \frac{3N}{2} \ln(pV^{5/3}).$$ (9.24)

(i) Draw a diagram of the Carnot cycle in the (p, V) plane, labeling the sides (A), (B), (C), and (D), and marking in p_1 to p_4 and V_1 to V_4.

(ii) Draw a diagram of the Carnot cycle in the (S, T) plane, labeling the sides as before, and marking in T_1, T_2, S_1 and S_2.

(iii) Use the equations for isothermal and adiabatic expansions to write down four equations, (one for each step), relating the ps, Vs and Ts together. Be very careful to get the subscripts right.

(iv) Use the above equations to derive the following relation between volume and temperature ratios,

$$\left(\frac{V_3}{V_2}\right) = \left(\frac{V_4}{V_1}\right) = \left(\frac{T_1}{T_2}\right)^{3/2}.$$ (9.25)

Hint: A useful trick to use is to write

$$p_2 V_2^{5/3} = p_2 V_2 V_2^{2/3} = N T_1 V_2^{2/3}.$$ (9.26)

(v) At which of the steps (A–D) does the system take in heat Q_1; at which step does it give up heat Q_2? Deduce equations for Q_1 and Q_2 in terms of T_1, T_2, S_1 and S_2. Deduce an equation for the work done by the system, W, and thus the efficiency η of the cycle. (η is defined as W/Q_1.)

(vi) Use (9.24) to calculate S_1 in terms of p_1 and V_1; also calculate S_2 in terms of p_2 and V_2. Then use these to derive a formula for $S_1 - S_2$ in terms of the ratio V_2/V_1 only.

(vii) Finally, use the results in (v) and (vi) to derive the formula for work done,

$$W = N(T_1 - T_2) \ln(V_2/V_1).$$ (9.27)

Note that the work done per cycle depends only upon the operating temperatures, T_1 and T_2, and the 'compression ratio' V_2/V_1.

(viii) I'm tooling down I-81 in my 5 liter Le Behemoth at 100 m.p.h.,† with the rev-meter sitting on 3600 r.p.m. Given that the cylinders work between 20°C and 1000°C, at a compression ratio of 10, and that 1 mole of air occupies 22.4 liters, what is the power ouput of the engine – assumed to be operating a perfect Carnot cycle. Given that 1 horsepower is 746 watts, express this in hp. Note that a real internal combustion engine operates a different type of cycle, an

† This is a *thought* experiment!

Otto cycle, so your result is just a ballpark figure. (The watt is the SI unit of rate of doing work (power), and is 1 joule/sec.)

Solution: (i) The isothermal parts of the cycle, (A) and (C) are given by pV =const, whilst the adiabatic parts, (B) and (D), are given by $pV^{5/3}$=const, so connecting the four sets of (p, V) points in the (p, V) plane up by such curves leads to Fig. 9.4

(ii) During the isothermal expansion the temperature remains constant at T_1 whilst the entropy increases from S_1 to S_2. We can see that entropy increases with volume for an isothermal expansion, by putting the ideal gas equation into the equation for entropy, giving

$$S = \text{const.} + \frac{3N}{2}\ln(pV^{5/3}) = \text{const.} + \frac{3N}{2}\ln(TV^{2/3}). \qquad (9.28)$$

Physically, the entropy increases because the gas has more states to occupy due to the increase in volume. During the adiabatic expansion, the entropy remains constant at S_2, whilst the temperature decreases to T_2. This can be seen from the entropy equation above. In physical terms, the temperature of the gas has to drop to supply the work necessary to expand the gas against an external pressure. The isothermal compression then decreases entropy from S_2 to S_1 at constant temperature T_2, and the adiabatic compression increases temperature from T_2 to T_1 at constant entropy S_1. This cycle in the (S, T) plane is shown in Fig. 9.5.

(iii) Using the equations for isothermal processes $pV = \text{const.}$, and for adiabatic processes $pV^{5/3} = const$, and the ideal gas law $pV = NT$, we can obtain the following four equations relating the ps and Vs at each

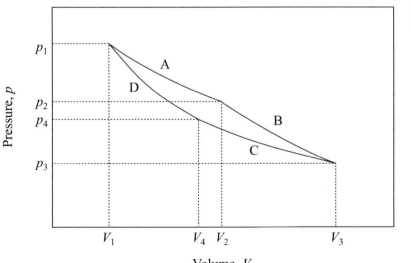

Fig. 9.4. Carnot cycle in the (p, V) plane.

step,

$$p_1 V_1 = p_2 V_2 = N T_1$$
$$p_2 V_2^{5/3} = p_3 V_3^{5/3}$$
$$p_3 V_3 = p_4 V_4 = N T_2 \qquad\qquad (9.29)$$
$$p_4 V_4^{5/3} = p_1 V_1^{5/3}.$$

(iv) If we use the hint given to write, $p_2 V_2^{5/3} = N T_1 V_2^{2/3}$ and $p_3 V_3^{5/3} = N T_2 V_2^{2/3}$, and then use the second of the equations in (9.29) to equate the two terms, we obtain,

$$T_1 V_2^{2/3} = T_2 V_3^{2/3}, \qquad\qquad (9.30)$$

from which we can derive,

$$\left(\frac{V_3}{V_2}\right) = \left(\frac{T_1}{T_2}\right)^{3/2}. \qquad\qquad (9.31)$$

Performing the same set of manipulations on the fourth equation in (9.29) produces the second part of the required equation, giving

$$\left(\frac{V_3}{V_2}\right) = \left(\frac{V_4}{V_1}\right) = \left(\frac{T_1}{T_2}\right)^{3/2}. \qquad\qquad (9.32)$$

(v) The system takes in heat Q_1 during the isothermal expansion (A), since the entropy increases at constant temperature, and heat is given by $\Delta Q = T\Delta S$. More physically, this heat is needed for the gas to be able to perform the work necessary to expand against an external pressure without its internal energy, and thus temperature, falling. Similarly, the system gives up heat Q_2 during the isothermal compression (C) for the opposite reasons. From the equation $\Delta Q = T\Delta S$ applied to (A) and (C) we get that,

$$Q_1 = T_1(S_2 - S_1) \qquad ; \qquad Q_2 = T_2(S_2 - S_1). \qquad (9.33)$$

Fig. 9.5. Carnot cycle in the T–S plane.

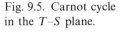

The work done by the system is given by $W = Q_1 - Q_2$, so that,

$$W = (T_1 - T_2)(S_2 - S_1),\tag{9.34}$$

and the efficiency, η, is

$$\eta = \frac{W}{Q_1} = 1 - \frac{T_2}{T_1}\tag{9.35}$$

exactly as in the reversible limit of (9.22). (vi) Since S_1 is the entropy at pressure p_1 and volume V_1, and S_2 the entropy at p_2 and V_2, we have that,

$$S_1 = \text{const.} + \frac{3N}{2}\ln(p_1 V_1^{5/3})$$

and

$$S_2 = \text{const.} + \frac{3N}{2}\ln(p_2 V_2^{5/3}).\tag{9.36}$$

Thus,

$$S_2 - S_1 = \frac{3N}{2}\ln(\frac{p_1 V_1^{5/3}}{p_2 V_2^{5/3}}),\tag{9.37}$$

and, using the first of the equations in (9.29) to eliminate $\frac{p_2}{p_1}$, we get,

$$S_2 - S_1 = N\ln\left(\frac{V_2}{V_1}\right).\tag{9.38}$$

(vii) Simply inserting the result from (vi) for $S_2 - S_1$ into the equation for ΔW in (v) gives us the result for the work done by the system,

$$\Delta W = N(T_1 - T_2)\ln\left(\frac{V_2}{V_1}\right).\tag{9.39}$$

(viii) We now approximate the Le Behemoth engine as operating an ideal Carnot cycle. The operating temperatures are given by, $T_1 = 1000°C$, $T_2 = 20°C$. Since temperature *differences* in Celsius and Kelvin are identical, $T_1 - T_2$ in energy units is $980k_B$ joules, where k_B is Boltzmann's constant given on p. 116. Now, V_2/V_1 is the compression ratio which is given as 10. Finally, since 1 mole of air occupies 22.4 liters, our 5 liter engine has $5/22.4 = 0.223$ moles of air in it, which means that $N = 0.223 \times N_{Av}$, where N_{Av} is Avogadro's number, also given on p. 116. Substituting this information into the work equation, we find that the work done per cycle is 4180 joules. Since the engine executes this cycle 3600 times per minute, which is 60 times per second, its power, which is work done per time, is $60 \times 4180 \approx 251\,000$ watts. This is $251\,000/746 = 335$ hp, which is a not unreasonable figure. A real engine is less that half as efficient as an ideal Carnot cycle, but has more moles of working substance, as the fuel which is injected and exploded is taken around the cycle along with the air.

(2) During some integral number of cycles, a reversible engine working between *three* heat reservoirs: absorbs Q_1 joules of heat from a reservoir at T_1 K, also absorbs Q_2 joules of heat from a reservoir at T_2 K, delivers Q_3 joules of heat to a reservoir at T_3 K, and performs W joules of mechanical work.

(i) Write equations – in terms of some or all of the quantities Q_1, Q_2, Q_3, T_1, T_2, T_3, and W – expressing the conditions imposed by the first and second laws of thermodynamics.

(ii) If $T_1 = 400$ K, $T_2 = 200$ K, $T_3 = 300$ K, $Q_1 = 1200$ joules, and $W = 200$ joules, find Q_2 and Q_3 in joules.

Solution:

(i) The first law requires that the energy inputs and outputs be equal

$$Q_1 + Q_2 = Q_3 + W. \tag{9.40}$$

The second law requires, since everything is reversible, that the total entropy be unchanged. The working substance has been taken through an integral number of cycles, so it is back where it started from as far as entropy is concerned. Thus one has only to consider the heat reservoirs, for which we know that $\Delta S = (Q/T)$. Equating entropy decreases and increases yields

$$\frac{Q_1}{T_1} + \frac{Q_2}{T_2} = \frac{Q_3}{T_3}. \tag{9.41}$$

(ii) If $T_1 = 400$ K, $T_2 = 200$ K, $T_3 = 300$ K, $Q_1 = 1200$ joules, and $W = 200$ joules, the above equations give:

$$Q_3 - Q_2 = 1000, \quad \text{and} \quad \frac{Q_3}{300} - \frac{Q_2}{200} = \frac{1200}{400} = 3$$

where Q_3 and Q_2 are in joules. The solution of these equations is $Q_3 = 1200$ joules, and $Q_2 = 200$ joules.

10

Fluctuations and the arrow of time

Time doth transfix the flourish set on youth,
And delves the parallels in beauty's brow,
Feeds on the rarities of Nature's truth,
And nothing sows but for his scythe to mow.

                                        Shakespeare

Isolated systems left to themselves, we have argued, evolve in such a way as to increase their entropy. The successes achieved by the use of this principle in the last two chapters cannot, however, hide the fact that it has up to now been pure assertion, more than reasonable to be sure but only vaguely connected to the motions of the microscopic constituents of matter. Let us now explore this connection, and the reasons why time and disorder flow in the same direction.

This is a curiously subtle question. From the point of view of common sense, it is not a puzzle at all. Shown a film of, for example, a lit match progressively returning to its unburnt condition, we instantly recognize a trick achieved by running a projector backwards. While it is by no means true that every macroscopically quiescent material object is in equilibrium,† it is a general feature of common experience that left to themselves inanimate isolated systems become increasingly disordered. Even when the process is very slow, it is inexorable: cars eventually rust away. More often, it happens in front of your eyes: the ice cube in your drink melts, the sugar you add to your tea dissolves and never reassembles in your spoon.

What, then, is the problem? It was alluded to in the second sentence of this chapter, and can be sharpened by asking a troublemaking question: Through what mechanism is the entropy maximized? This general and seemingly simple question has no general and simple answer, and depending on your point of view, the last two chapters have meretriciously evaded or astutely avoided it. This author leans towards the latter view. As we have seen, the bold hypothesis that all states of the same energy are equally likely in equilibrium has

---

† Ordinary glass, as a prime example, is in a *very* long-lived but non-equilibrium state achieved by rapid cooling of a suitable liquid. If held at a temperature somewhat below the melting point for long enough, glass becomes an ordered solid and in this condition of true equilibrium loses the technologically useful property of gradually softening when heated.

given deep insights – e.g. into the meaning of temperature, – and allowed unexpected predictions – e.g. the distribution of velocities of the molecules in a gas in equilibrium. For the properties of systems in equilibrium, Statistical Mechanics with the caveats about what states of motion are to be included – gas atoms confined by container walls – is certainly right, and unbeatable as a calculational method.

That it is, nonetheless, not completely frivolous to question the foundations of the second law (9.11) can be seen by invoking something that physicists call Time Reversal Invariance, but which is perhaps better described as the principle of the reversibility of motion. Every microscopic theory of motion we have, classical as well as quantum mechanical, permits time evolution to occur in a sequence precisely opposite to a given one: if an apple falls, accelerating as it does so, it is perfectly possible to imagine an apple projected upwards, decelerating as it rises. In short, a film run backwards of the apple falling shows a sequence of events allowed by the Laws of Motion. In fact, such a movie, at least superficially, does not violate our common sense. The problem, which only arises when many particles are involved,† is that motion reversibility applies also to the microscopic constituents of a system; so that a backward running movie, even when it obviously violates common sense, e.g. of ice-cubes spontaneously growing in a glass of soda in your living room, shows movements allowed in principle by the laws of motion. And that is the apparent paradox.

As already mentioned in the last chapter, there is nothing in this paradox that violates the principle of the conservation of energy. Energy conservation alone does not exclude the evolution of order from disorder, however counter-intuitive that might be. And counter-intuitive it is indeed. Imagine a poker and room at the same temperature. The poker would spontaneously become hotter than the room only if every microscopic constituent of the system of room plus poker underwent motions time reversed from those that occur when a hot poker cools. A typical process in the latter case is a rapidly moving iron atom on the surface of the poker colliding with a gas atom in the surrounding air and knocking it into a state of high velocity. This atom would then collide with the typically slower other gas atoms, and thus contribute to the gradual raising of the temperature of the air. In the time reversed case, a rare (because it must be in the low-probability

---

† This point is worth expanding on. The motion of a single ball colliding with the edge of a billiard table can be reversed without seeming unlikely, provided that the observation is not careful enough to spot the slowing down due to friction. On the other hand, the direction of time is obvious in the 'break' at the start of a game of pool, when a triangle of 15 balls is struck by an incoming one.

tail of the Maxwell distribution) very rapidly moving gas atom would collide with a rare slowly moving iron atom and transfer energy from the air to the poker. Such processes indeed must occur as a poker cools, but for them to heat up a poker significantly is overwhelmingly unlikely. Put another way, since the hot poker and cold room have a (much) lower entropy than the cool poker and slightly warmer room in equilibrium, there are many fewer microscopic states of motion consistent with the initial non-equilibrium macroscopic state than with the final equilibrium one. Almost all of the former microstates are 'cooling,' i.e. when developed forward in time under the influence of the laws of motion they lead to situations in which some gas atoms are moving faster than before and some iron atoms slower. For each 'cooling' state there is an exactly time reversed 'conspiratorial' one which concentrates its energy into a few rapidly moving atoms and then transfers energy to the iron atoms in the poker. The 'cooling' and the 'conspiratorial' states have the same energy and both occur as possible states of motion in the final equilibrium of the isolated system of poker and room. However, they form a truly insignificant fraction of all the microstates consistent with that equilibrium. The conspiratorial states are very precisely aimed: disturb one atom and the conspiracy goes awry; the heating stops and cooling again ensues. The increase of disorder, one would then say, is associated with the time evolution of an overwhelmingly large fraction of the microscopic configurations consistent with the gross state of the matter under consideration.

For most purposes, and indeed for most practical calculations, that is enough. At a conceptual level, however, the arguments of the last paragraph have not disposed of the problem raised by time reversal invariance, as can be seen by the following. We saw that given the fact of a hot poker – of which I'm getting tired, and maybe you are too, so let's suppose it's a steam-iron, turned off, from now on – and a cold room, the overwhelmingly likely forward evolution in time is towards equilibrium. Now, as just emphasized, every one of these evolutions has its time reversed counterpart, i.e. a movie of the microscopic motions run backwards. Thus, *given* that at some time the iron is hot and the room cold, and *assuming* that it got hot under the influence of the same laws of motion that will cool it, it is overwhelmingly likely that these 'conspiratorial' motions were responsible. So time reversal says that a spontaneous entropy reducing sequence of collisions – a giant 'fluctuation' – has no sense of time: it arose in the past in the same way as it decays into the future. Whoa, you say. That's nonsense. That iron didn't get hot all by itself. Even if I wasn't here, I *know* that it must have been turned on. Of course, you're right. But, you see, probability and mechanics in and of themselves have nothing to say about the direction of time, if the underlying laws are invariant under motion reversal.

So, how is it that you 'know' that that iron had been heated electrically? The point is that in the world we live in it is a lot easier to heat a subsystem, or, more generally, to achieve a non-equilibrium initial condition by exploiting engines or concentrated sources of energy made by humans or plant-life, and all depending ultimately on sunlight, than to wait for an exceedingly rare fluctuation. And, strangely, that seems to be the origin of the direction of time: that we exist in a universe, not by any means in equilibrium, in which stars shine.

This breath-taking conclusion is worth bringing home in another way. Consider an ordinary deck of 52 playing cards. If it were shuffled randomly, which is actually not so easy to do, there would be a truly minuscule but finite probability of achieving any specified order, e.g. the order in which a new deck is packaged.† *If* – and this is the enormous if – the special ordering was in fact achieved by random shuffling, then there is a probability of essentially unity that a few shuffles later (*and* a few shuffles earlier) the deck will be (or was) not so ordered. This is the essence of the symmetry in time of a giant fluctuation. On the other hand, if one picks up a pack of cards and finds it ordered, the most sensible hypothesis by far is that someone did it. When the ordered deck is then treated as a prepared starting condition, the past has to be considered manipulated and having nothing in common with random shuffling: the disordering only occurs in the future and the chance of perfect order re-occurring without outside intervention is to all intents and purposes zero.

There is a logically possible hypothesis about our part of the universe which would explain why the energy in our vicinity is very far from being randomly distributed among the microscopic constituents of matter. Could our immediate celestial environment be a giant fluctuation in an isolated universe in equilibrium as a whole? The answer is that neither astronomical observations nor currently accepted theories of cosmology give any support to the view that our part of the cosmos is in any way special.

The rest of this chapter deals with the approach to equilibrium of a simple model. As in previous chapters, I hope to demystify some of the thoughts in the previous paragraphs by making them concrete.

† Equal to $\frac{1}{52!} \approx 1.2 \times 10^{-68}$.

Fluctuations and the approach to equilibrium: Ehrenfest's dogs

Dealing with the evolution in time of a reasonably realistic statistical system is a technically difficult problem. Even for an ideal gas – the word ideal implying, as you know, a density so low that collisions between molecules are rare – collisions cannot be ignored, because they are, in fact, the instruments that shuffle any particular molecule between its states of motion. The approach to equilibrium thus depends on the details of the motion, and in this sense is a less general phenomenon than equilibrium *per se*. In order not to lose the woods for the trees, it is necessary to look for simple but illustrative examples. Since we learned so much in the early part of this book from coin tossing, I am going to spend the rest of this chapter discussing the approach to equilibrium of a collection of 'two-level'systems, i.e. systems described, like coins, by two possible outcomes when their state of motion is investigated.

As a physical realization of this model, one could consider a collection of weakly interacting magnets each of which has two possible states, north pole up or down. A more whimsical illustration is based on one proposed by the Dutch physicists Paul and Tatiana Ehrenfest in 1907: let the 'subsystem' be fleas, and the 'states' residence on dog A or dog B, which I shall henceforth call Anik and Burnside, sleeping side by side. In order to simulate molecular agitation we shall suppose that the fleas jump back and forth between the dogs. Now we need something that plays the role of the rest of the system. Let us suppose that these are trained circus fleas, each equipped with a number, that jump when their number is called. The 'environment' agitating the fleas, which is like a reservoir of the last chapter, will here be something that calls out numbers at random, and our closed system will be the fleas and the reservoir.

In Chapter 8, we said that probability could be introduced into a discussion of equilibrium in two ways. One could either contemplate the evolution in time of a single subsystem, or think about many (an ensemble) of subsystems at one instant. In fact we used the second method to deduce the Gibbs ensemble. Now we are considering a subsystem out of equilibrium, and even the second method requires the explicit consideration of time. In many ways it is more instructive to use the first method, and we shall do that first.

The subsystem in our whimsical example is specified by deciding on a certain number of fleas, which number I shall take to be 50. Even with such a small number there immediately arises a difficulty with keeping track of possibilities. There are 2^{50} ways in which the fleas can be distributed among two dogs – approximately 10^{15}, a thousand million million! This large number has come about because we have treated the fleas as 'distinguishable.' In the experiment of tossing a coin 50 times, the same number would occur if one kept track of the order in which heads or tails came up. As in that problem we can simplify matters by only recording how many, and not which, fleas are on Anik. Then, since the range is $0-50$, there are only 51 possibilities. By not paying attention to which fleas are on Anik, we are as it were only recording the 'macrostate,' and this is a reasonable thing to do in discussing the approach to equilibrium which typically is a perceptible process.

To simulate the approach to equilibrium it is necessary to start from a configuration which occurs almost never in the maximally random state of affairs. Let us suppose that in the beginning Anik has no fleas at all. Now we agitate the fleas by arranging to have a computer produce numbers at random between 1 and 50,† and to transfer the flea with the called number from one dog to the other. What transpires is totally unpredictable at every step and yet, to those informed about statistical matters, totally understandable.

A typical sequence is shown in Figs. 10.1 and 10.2. In the first

Fig. 10.1. Number of fleas on Anik due to random jumps: early time development.

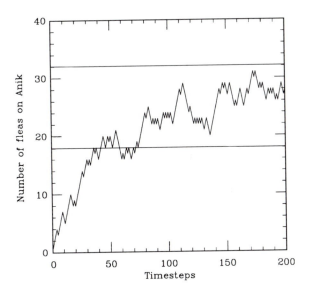

† This may seem like a contradiction in terms, because the basic elements of a computer merely carry out simple well defined mathematical operations. In the next chapter we shall encounter an example of just how one programs a computer to generate random numbers.

step it is certain that the number called will belong to a flea on Burnside. In the second step the probability of this happening again is $\frac{49}{50}$. Thus, in the beginning there is what seems like a steady march towards an equal partition of the fleas between the dogs. The early time development is shown in Fig. 10.1. After something like 50 steps we get to a situation where sometimes Anik and sometimes Burnside has more fleas. In this region one would expect that every one of the 2^{50} configurations mentioned above would be equally likely. This translates, exactly as in Chapter 3, into a binomial distribution with $p = \frac{1}{2}$, i.e. the repeatable random event of tossing 50 fair coins. An examination of Fig. 10.2 confirms this expectation. The horizontal lines have been drawn to include 2 standard deviations on either side of the mean, which would give the 95% range for a normal distribution and approximately that for a binomial. [The standard deviation here is (Chapter 3) $\sqrt{50}/2 \approx 3.5$.] Your eye tells you that, except for the initial transient, fluctuations outside this range are indeed rare.

We can be more quantitative. Fig. 10.3 is a histogram of the relative durations of the possible outcomes, constructed from Fig. 10.2 with the first 100 steps omitted. Superimposed on the histogram is the binomial distribution. The agreement is very good.† The dance of the fleas in Fig. 10.2 has thus very quickly forgotten its unusual

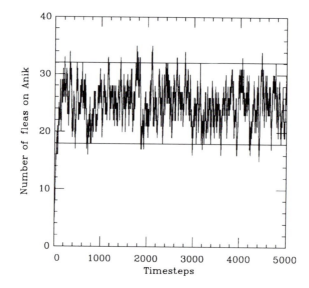

Fig. 10.2. Long time behavior of fleas on Anik showing fluctuations in equilibrium.

---

† But not perfect. How reasonable the disagreement is, given the number of trials, is a question I am going to desist from asking, because ultimately that would include among other things a test of how well I have instructed my computer to generate random numbers.

starting point and become the endless jitterbug of 'equilibrium,' in which something as unlikely as a flealess dog simply never happens again without outside intervention.

Figure 10.2 also illustrates the previous remarks about motion reversal. Our model is in fact 'time reversal invariant' – because a string of random numbers in reverse order is just as random. After the first 100 or so steps, the evolution has no sense of time. If we were to expand the region near one of the reasonably large fluctuations away from the mean we would find that there is no characteristic feature of the build-up preceding the maximum deviation to distinguish it from the time reverse of the decay following the maximum. There is also no conclusive logical argument to rule out the possibility that the start of the trace shown in Fig. 10.1 has captured a truly giant fluctuation midway. Of course you and I know that that is not how the figure was produced.

In fact it is virtually impossible to wait long enough for the initial configuration in Figs. 10.1 and 10.2 to occur as a fluctuation in equilibrium, where it has a probability of 2^{-50}. To have a reasonable chance of witnessing such a fluctuation one would have to allow a number of timesteps approximately equal to the reciprocal of this probability, the previously mentioned thousand million million, to elapse. Thus, to reach the unlikely configuration of a totally clean Anik by random shuffling of fleas between equally dirty dogs, even with this *very* small system of 50 'fleas,' one would need a trace roughly two hundred thou-

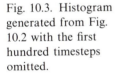

Fig. 10.3. Histogram generated from Fig. 10.2 with the first hundred timesteps omitted.

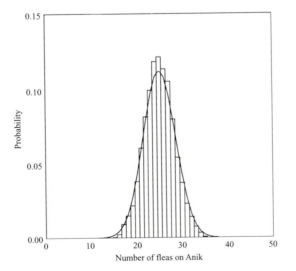

sand million times as long as Fig. 10.2, which extends for 5000 steps. Since Fig. 10.2 is about 5 cm wide, the length of the required trace would be some ten million kilometers – and the distance to the moon is only about four hundred thousand kilometers. The Law of Large Numbers is at work here making an unlikely event overwhelmingly unlikely, and the timereversal paradox academic. Though not logically certain, it is roughly 99.999999999999999% probable that in Fig. 10.1 time is running in the direction of increasing disorder.

Entropy

Ehrenfests' dogs bring into focus the essential characteristics of time's arrow. (i) A starting point macroscopically distinct from equilibrium is overwhelmingly likely to evolve to greater disorder, i.e. towards equilibrium. (ii) In equilibrium, fluctuations have no sense of time. (iii) Giant fluctuations from equilibrium to extremely unlikely states are extremely rare. Even for the rather small system we considered, these uses of the words 'overwhelmingly' and 'extremely' are *very* conservative.

The word entropy did not occur in the last section. As a matter of fact, there is more than one way to introduce that notion here. One way to proceed is to note that the states with n fleas on Anik are 'macrostates,' each allowing for $\binom{50}{n}$ assignments of distinct fleas or 'microstates.' One could then simply call the logarithm of the latter number, the entropy, i.e. let $S(n) = \ln \binom{50}{n}$. Now, we know (from Pascal's triangle if you wish) that these combinatorial coefficients have a maximum half way, at $n = 25$, and get smaller in either direction. Fig. 10.2 could be translated into a diagram of entropy vs. timesteps via this rule. Like Fig. 10.2, it would start from zero because $S(0) = \ln \binom{50}{0} = \ln 1 = 0$, and also like Fig. 10.2 it would have spiky fluctuations, but now – because the logarithm is a smoothly increasing function of its argument – downwards on both sides of maxima occurring when $n = 25$. The entropy introduced in this paragraph is sometimes called the Boltzmann entropy. Although it fluctuates in equilibrium, these fluctuations diminish as the size of the system increases for reasons discussed in Chapter 3. This entropy can be associated with a *single*† time trace like Fig. 10.2. If (and only if) a system is sufficiently large, fluctuations become insignificant, and the Boltzmann entropy increases steadily as equilibrium is approached.

---

† Some people, who have thought deeply about these questions find this deeply satisfying.

It is closer to the spirit of Chapter 8, where probabilities were introduced by contemplating many subsystems, to associate a constant entropy with equilibrium. I shall call this the Gibbs or statistical entropy. From this point of view, the distribution in Fig. 10.3, which uses *all* the information except for the first 100 steps of the time trace in Fig. 10.2, determines an entropy. The general expression for entropy is given by (8.12), and we can easily enough work that out. It is easier, however, to interpret the result when the outcomes are counted distinguishing the fleas, because in this case the maximally random situation corresponds to the 2^{50} equally likely outcomes already mentioned, for which the entropy is $50 \ln 2 = 34.657$. If we subdivide the probability for m fleas on Anik into the $\binom{50}{m}$ states corresponding to distinguishing which fleas are on each dog and follow the logic of the argument above (8.8) to add in the entropy associated with this finer subdivision, we arrive at the expression

$$S = - \sum_{m=0}^{50} P_m \ln P_m + \sum_{m=0}^{50} P_m \ln \binom{50}{m}. \tag{10.1}$$

When I tell the computer to calculate this for the empirically determined distribution in Fig. 10.3 the answer is 34.651, which differs slightly from $50 \ln 2$, presumably because of numerical errors, but is perhaps close enough to convey to perceptive observers like you and me that the system is maximally disordered.

Gibbs entropy in a non-equilibrium context

To associate a Gibbs or statistical entropy with the early, and consequently non-equilibrium, part of the time development we have been discussing we need more information than the single time trace we have been examining in such detail. Since statistical entropy is a property of a distribution, we need to assign probabilities to every timestep of the process, which means that we have to contemplate an ensemble of subsystems and define probabilities in terms of occurrences in the ensemble. One way to proceed would be to create a very large number of traces like the one in Fig. 10.1, all of them starting with the same configuration. Since the sequence of random numbers would be different in each run, these traces would progressively differ from one another. At any given time step one could then calculate a histogram like Fig. 10.3. To give reliable distributions, this method would require a very large number of runs. Fortunately, there is a much simpler way of implementing the idea, which does not require a random number generator. Let us try to calculate a distribution function $P(m)$, with m running from 0 to 50, which changes from step to step, and reflects the random transfer of fleas from dog to dog. At the start of the process we are told with certainty that there are no fleas on Anik. In the language of probability this just means that at timestep 0, the distribution is: $P_0(0) = 1, P_0(1) = P_0(2) = \cdots = P_0(50) = 0$. Now we can argue that the distribution at timestep t determines the distribution at the next timestep $t + 1$. [To indicate which time is being referred to, we need another label, which we shall write as a subscript.] The assumption that the fleas are being called at random implies that

$$P_{t+1}(m) = \frac{m+1}{50} P_t(m+1) \ + \ \frac{51-m}{50} P_t(m-1). \qquad (10.2)$$

This is understood by saying it in words. Anik can have m fleas at timestep $t + 1$ *either* because she had $m + 1$ at step t, and one jumped off, which has a probability proportional to $m + 1$, *or* because she had $m - 1$ and one jumped on, which has a probability proportional to the $51 - m$ fleas that were on Burnside at step t.

A computer can be programmed to develop the distribution corresponding to the initial certainty forward in time using (10.2). There is, however, one artificiality in this time evolution: at odd (even) timesteps only odd (even) numbers of fleas can be on Anik. This can be remedied by averaging (10.2) over two forward steps. The resulting evolution is shown in Fig. 10.4.

This three dimensional plot is obtained by stacking together the distributions at successive times. It shows very clearly the initial certainty evolving to a distribution (which, not surprisingly, can be shown to be a binomial) in which the outer regions in the range of possibilities are extremely unlikely. Now, at each step one can indeed calculate an entropy, using (10.1). The result, Fig. 10.5, shows that the entropy rises steadily from zero to $50 \ln 2$.

If a picture is worth a thousand words, the figures in this section are worth puzzling over, because they do capture the process of disorder evolving from order, albeit in a particularly simple system.

Summary and comments

Although we have learned nothing here about real dogs or fleas, and although it may not impress you at all to learn that the model is actually rather a good description of the decay of nuclear magnetization – which magnetization is detected in M(agnetic) R(esonance) I(maging) – the lessons of the last sections are general. We have constructed a model that evolves in time from a prepared initial state towards and to equilibrium. Our system passes randomly through all possible configurations, with the consequence that any particular configuration of labeled fleas, like the initial condition, re-occurs very rarely. Thus the evolution is to the likely from the unlikely. By contemplating many such evolutions from the same initial state, we have been able

Fig. 10.4.
Probabilities of fleas on Anik with the passage of time.

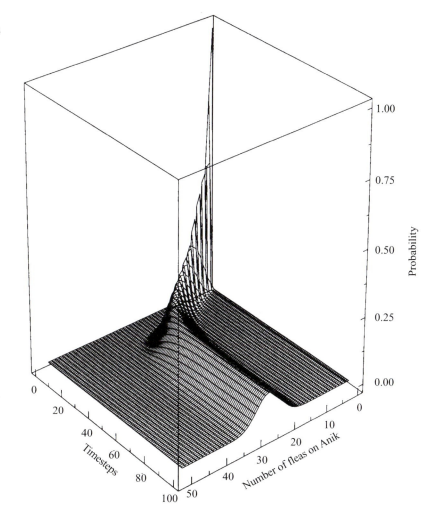

to associate an entropy with non-equilibrium fluctuations, and thereby to follow the progress from low to high entropy.

The rate at which equilibrium was reached depended very particularly on the hypothesized mechanism controlling the evolution. This feature is general: in any physical situation, rates of approach to equilibrium can only be understood when the corresponding dynamics have been identified and carefully studied.

If our system has reached equilibrium, what is its temperature? This is a very instructive question. Equations. (9.8) and (9.14) show that when no work is done by a system, so that the second term in the former equation is zero, one has

$$T = \frac{1}{\beta} = \frac{\Delta E}{\Delta S}. \tag{10.3}$$

In words, Temperature is the rate of change of Energy with Entropy in equilibrium when no work is done. But the concept of energy has not entered our model. In the only equilibrium state there is, changing the energy changes the entropy not at all. Thus, the temperature would have to be assigned the value infinity here, though it probably makes more sense to say that it has no very useful meaning in this context. The model could be generalized by assigning a penalty in energy every time a flea jumped on to the initially flealess dog, and

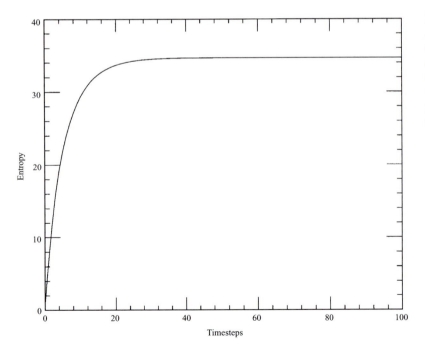

Timesteps

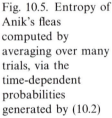

Fig. 10.5. Entropy of Anik's fleas computed by averaging over many trials, via the time-dependent probabilities generated by (10.2)

then a temperature could be sensibly defined. [See solved problem (2) at the end of this chapter.]

It will be apparent that the model side-steps some of the questions raised in the first section of this chapter. In particular, no attempt has been made to understand how the dynamics of the combination of environment and subsystem appears to the latter as a random number generator. Some of these questions are addressed in the next chapter.

More by accident than design I have to confess, the ideas of Chapters 2 and 3 have been used here in precisely the form in which they were there developed. A way of thinking born in the gambling salon has indeed cast light on some of the secrets of the inner workings of nature.

Solved problems

(1) <u>Steady-state distribution of fleas on Ehrenfest's dogs.</u> Prove the remark made in the second paragraph below (10.2) that under the influence of this equation the distribution describing the number of fleas on Anik evolves to the binomial:

$$P(m) = \binom{50}{m} (\frac{1}{2})^{50}. \tag{10.4}$$

Show this by substituting the above form on the right hand side of (10.2) and observing that it reproduces itself on the left, thereby demonstrating that it is the steady-state of the time evolution. It is also the end product of the evolution. Are you surprised?

Solution: As suggested, we substitute $P(m)$ into both sides of

$$P_{t+1}(m) = \frac{m+1}{50} P_t(m+1) + \frac{51-m}{50} P_t(m-1), \tag{10.2}$$

and verify that equality holds. This requires, after cancelling out a factor of $(\frac{1}{2})^{50}$, that,

$$\binom{50}{m} = \frac{m+1}{50} \binom{50}{m+1} + \frac{51-m}{50} \binom{50}{m-1}. \tag{10.5}$$

So the problem reduces to proving an algebraic relation among binomial coefficients. Noting that,

$$\frac{m+1}{50} \binom{50}{m+1} = \frac{m+1}{50} \frac{50!}{(m+1)!(49-m)!}$$
$$= \frac{49!}{m!(49-m)!} = \binom{49}{m}, \tag{10.6}$$

with a similar formula for the second term on the right of (10.5), one sees that (10.5) reduces to

$$\binom{50}{m} = \binom{49}{m} + \binom{49}{m-1}. \tag{10.7}$$

But this is an example of a relation that was proved in solved problem (1) of Chapter 3. We have thus demonstrated that the binomial distribution is a steady-state – i.e. not varying with time – solution of the time evolution equation. This is not surprising: the evolution randomizes the location of the fleas, with the result that the probability of a certain number of fleas on Anik is equal to the probability of obtaining that same number of heads in 50 tosses of a fair coin.

(2) <u>Temperature in the Ehrenfest Model.</u> Suppose that Anik is cleaner than Burnside, providing a less friendly environment for fleas. Model this by assuming an energy cost ϵ to be paid by a flea jumping from Burnside to Anik. Let the fleas be at an effective temperature, in energy units, T, and define $\Delta = \epsilon/T$. Attempt an argument for why (10.2) should now be changed to:

$$P_{t+1}(m) = \frac{m+1}{50}P_t(m+1) + \frac{50-m}{50}\left[1 - e^{-\Delta}\right]P_t(m)$$
$$+ \frac{51-m}{50}e^{-\Delta}P_t(m-1). \tag{10.8}$$

What would you guess would be the equilibrium distribution now?
Solution: Under the new conditions, we expect that in equilibrium any particular flea will spend more time on (dirty) Burnside than on (clean) Anik. This is implemented in (10.8), which (as you will verify by examining the three terms on the right hand side) imposes the rules that when the number of one of the fleas on Anik is called it jumps to Burnside with probability unity (term 1), but if one of the fleas on Burnside is called it either stays put with probability $1 - e^{-\Delta}$ (term 2), or jumps with probability $e^{-\Delta}$ (term 3). By making the jump-probability from Burnside to Anik smaller than the reverse process by the Gibbs factor $e^{-\Delta}$ equilibrium is achieved in the steady state, as we shall now see. If the end product of the time evolution for any particular flea is a Gibbs distribution, it will be that corresponding to a system with two levels, discussed in solved problem (4) at the end of Chapter 8. In short, the probability p of a flea being on Anik, and the probability $1 - p$ of one being on Burnside should be given by

$$p = \frac{e^{-\Delta}}{1 + e^{-\Delta}}, \text{ and, correspondingly, } 1 - p = \frac{1}{1 + e^{-\Delta}}. \tag{10.9}$$

[Note that p is less than $1 - p$, Anik being a less desirable place for fleas.] For 50 fleas the equilibrium probability distribution would then be the binomial corresponding to 50 tosses of an unfair coin with outcome probabilities p and $1 - p$:

$$P_{eq}(m) = \binom{50}{m}p^m(1 - p)^{50-m}. \tag{10.10}$$

It can be verified via steps similar to, but slightly more involved than, those in the last solved problem that (10.10) and (10.9) indeed are a stationary solution of (10.8). In short, the effective temperature of the fleas determines how many are willing to put up with Anik's cleanliness. In the high temperature limit, $\Delta \ll 1$, the result of Problem (1) is recovered. At very low temperatures, $\Delta \gg 1$ few fleas leave the snug comfort of Burnside.

11

Chaos

HANNAH: The weather is fairly predictable in the Sahara
VALENTINE: The scale is different but the graph goes up and
down the same way. Six thousand years in the Sahara looks
like six months in Manchester, I bet you.

from *Arcadia* by Tom Stoppard

It is time to correct and soften a striking difference in emphasis
between the chapters on Mechanics (6) and Statistical Mechanics
(8). Contrast the simple and regular motions in the former with the
unpredictable jigglings in the latter. The message here will be that even
in mechanics easy predictability is not by any means universal, and
generally found only in carefully chosen simple examples. Sensitivity
to precise 'aiming' *is* found in problems just slightly more complicated
than those we considered in our discussion of mechanics. However,
the time it takes for such sensitivity to manifest itself in perceptibly
different trajectories depends on the particular situation. 'Chaotic'
as a technical term has come to refer to trajectories with a sensitive
dependence on initial conditions, and the general subject of their study
is now called Chaos.

There is a long tradition of teaching mechanics via simple examples,
such as the approximately circular motion of the moon under the
gravitational influence of the earth. This is a special case of the 'two
body problem' – two massive objects orbiting about each other –
whose solution in terms of elliptical trajectories is one of the triumphs
of Newton's Laws. Elliptical trajectories describe the journey of the
earth around the sun and of the moon around the earth, but only
when all forces except the mutual attraction of the two bodies in
question are ignored. At this level of approximation, the orbits trace
and retrace closed curves. The next simplest case is the 'three body
problem' – sun, earth, and moon, for example – taking into account
the attraction between each body and the other two. No simple
solution to this problem exists, and it is not studied in the physics
curriculum. It is, however, of immense importance to astronomers,
and entire chapters on this so-called 'lunar problem' are contained
in textbooks on celestial mechanics. These discussions appear dense

with formulas even to someone accustomed to such things because they describe calculations of deviations of the moon's orbit from a simple curve, as a series in powers of small parameters. I have in front of me as I write a copy of a memoir entitled 'Researches in Lunar Theory,' by the distinguished American theoretical astronomer and mathematician George William Hill (1838–1914). Hill studied a simplified model in which the mass of the moon is treated as arbitrarily small, so that it responds to but is assumed not to influence the motion of earth and sun. His seminal work, dated 1877, caught the attention of the most eminent French mathematician of his time, Henri Poincaré (1854–1912).† Poincaré's researches are to be found in his monumental three volume *Les méthodes nouvelles de la mécanique céleste* (1892–1899) Very close to the end of this opus, he shows that under certain circumstances the moon in Hill's model can develop a convoluted and complex trajectory 'that I am not even attempting to draw.' This is the first description of what today would be called a chaotic orbit. In his book *Does God Play Dice?*, Ian Stewart describes the motion in the following way:

> It's a bit like a bus which tours a city, repeatedly passing through the central square, but each time choosing at random from a million different bus-stops in the square itself. You can see the bus coming round again, and you know it will stop in the square – but you've got no idea at all which stop to wait at.

Notice that the word 'random' occurs in this description of a perfectly deterministic system. That moon-bus is a Newtonian bus, busily solving Newton's equations on its rounds. Given its precise position and velocity at one moment, and all the forces acting on it then and thereafter, there can be no doubt about where it will be. The key word here is *precise*. With an orbit of this kind imprecise initial conditions eventually lead to a complete loss of predictability, in the same way as the inevitable imprecision in the toss of a fair coin leads to an equal likelihood of heads or tails.

In Fig. 11.1 is shown a construction (related to a more technical one invented by Poincaré) which quickly identifies chaotic orbits. On a plane which intersects the trajectory are marked the crossing points of the orbit into the plane. An elliptical ('Keplerian') orbit always

---

† Poincaré emerges in the collected works of Hill (1905) as the author of a laudatory introduction in which he says of Hill's researches on the moon: '... he has been not only a gifted artist and a curious researcher, but also an original and deep inventor.' Poincaré was one of Hill's early admirers.

intersects the plane at the very same point. More general but still regular orbits are called 'periodic' if they intersect the plane at several discrete points, and 'quasi-periodic' if the intersections are along a closed curve which is stepped out in a regular way. A chaotic orbit – like the 'bus' – stipples the plane with what appears to be a random series of dots, even though each one is completely determined by the position and velocity at the previous intersection.

An evident problem with such an orbit is that unless it can be calculated precisely even a perfect knowledge of 'initial conditions' quickly fails to determine future motion. To use the bus analogy, even if one knew the present stop and could calculate the orbit, an imprecise calculation would get the bus-stop wrong after a few tours, and thereafter would be of absolutely no use as a predictor. One is thus led to ask: Can the calculation of planetary orbits be done, even in principle, with perfect precision? When these calculations are given

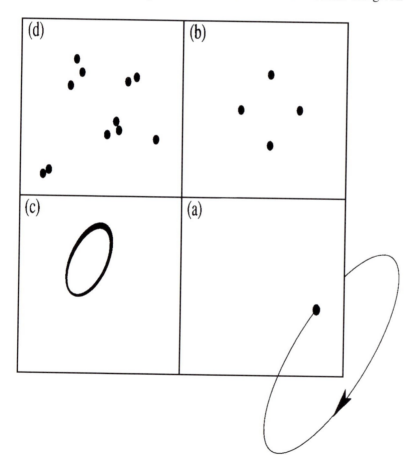

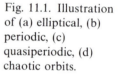

Fig. 11.1. Illustration of (a) elliptical, (b) periodic, (c) quasiperiodic, (d) chaotic orbits.

as series, we have to know if the addition of terms of a series can be carried to arbitrary precision. Here the notion of a convergent series is needed. To make this idea specific, consider the exponential series (5.10). It can be shown that the function e^x (see Fig. 5.2) is approached as precisely as one wishes in any finite region of x, by the sum of a sufficient number of terms of (5.10). The series is then said to be uniformly convergent, the qualifier indicating that it is possible to specify a required precision over the whole region.

Interestingly, Hill's memoir contains the remark, 'I regret that, on account of the difficulty of the subject and the length of the investigation it seems to require, I have been obliged to pass over the important question of the limits between which the series are convergent...' The passage from which this excerpt is taken shows that Hill was very well aware of the major question underlying the calculation of precise astronomical orbits.

Hill's professional interests were in the preparation of stellar maps as aids to navigation; his concern was not merely about mathematical niceties. As we have noted, his question is central to long term prediction. It also bears on a deeply interesting more fundamental one. Is our solar system stable? General considerations, such as conservation of energy, will not suffice to address this last issue, since in principle one of the planets could over eons work itself free from the pull of the sun, acquiring additional kinetic energy at the expense of other planets. Now, only if the series mentioned in the last paragraph converge uniformly can trajectories be calculated with arbitrary accuracy.

As the quotation from Hill's memoir indicates, this question was lurking in the background during much of the early history of theoretical astronomy. Before the 1880s various eminent mathematicians – e.g. P. G. L. Dirichlet (1805–1859) in 1858 – had asserted that they could prove the stability of the solar system, but there was nothing in writing. It was in this context that in 1885 King Oscar II of Sweden offered a prize for the answer to the question: 'For an arbitrary system of mass points which attract each other according to Newton's laws assuming that no two points ever collide, give the coordinates of the individual points for all time as the sum of a uniformly convergent series whose terms are made up of known functions.'

Poincaré was awarded the prize for a memoir in which he arrived at the distressing conclusion that such a solution is not possible in general. Roughly speaking, his negative result follows from the possibility that the periods of two planets may be *rationally related*, which means that the time taken for one planet to complete an integer number of orbits is equal to the time taken for another planet to complete the same or another integer number of its own orbits. When such a condition occurs the transfer of energy between even very weakly interacting planets is enhanced, leading to instabilities. As a example of such an energy transfer, consider two pendulum clocks hung close by on the same wall. Although the pendula are each beating time with a period of one second, there is no reason for them to be swinging in unison, i.e. to have their maximum displacements simultaneously. But that is in fact what they are very likely to be doing if they have been undisturbed for a long time, because of the phenomenon just described (called non-linear 'resonance') induced by the very weak coupling of the clocks via vibrations transmitted along the wall.

Further discussion of periodic versus chaotic orbits in the context of mechanics requires concepts rather different from those developed in this book, and quickly gets into the distinction between rational numbers (which can be written as a ratio of integers) and irrational ones (which cannot). [See the solved problems for some insight into this distinction.] One interesting piece of evidence that simple period ratios invite instability comes from the observation of gaps in the belt of variously sized small planets ('asteroids') which orbit the sun between Mars and Jupiter. The empty regions are found to occur where asteroids would have periods rationally related to Jupiter's. There is a particularly noticeable gap where the period ratio is 1 (asteroid) to 3 (Jupiter). Computations suggest that any asteroids once in this region would have had chaotic orbits which occasionally made large excursions putting them on a course close to Mars where its gravitational force swept them into outer space. As Ian Stewart engagingly puts it in the book just referred to: 'What Jupiter does is to create the resonance that causes the asteroid to become a Mars crosser; then Mars kicks it away into the cold and dark. Jupiter creates the opening; Mars scores.'

Unfortunately, this is not the prototype of a general explanation of structure in the asteroid belt. The details matter; they can only be revealed by long and careful computer calculations.

Chaotic orbits thus do occur in the mechanics of just a few bodies. But there are also stable and regular motions in such systems. In some models of weakly interacting systems the regular orbits can be studied, and it sometimes happens that the majority of orbits are regular – because, in a nutshell, there are many irrational numbers. However, our solar system is too complicated for these ideas to apply to it, and the question of whether it is stable has to be considered open.

In fact there is evidence that some parts of the system – the orbit of the planet Neptune, the tumbling in space of the moon Hyperion of Saturn – are chaotic.† A consensus seems to be emerging that in complicated dynamical systems chaos is just a matter of time: wait long enough and something irregular will show up. Luckily for us planetary motions as we know them have, apart from catastrophic collisions,‡ been steady enough on the scale of the evolution of life. Volcanos, earthquakes, floods, and environmental pollution are a bigger cause for concern.

How does chaos fit into the topics discussed so far in this book? Since much information about a trajectory that randomly stipples a surface as in Fig. 11.1 is conveyed by giving the probability for a dot to occur in different regions of the plane, we learn by extension that describing a many particle system by probabilities is quite natural. There is, however, an important difference. A chaotic orbit has no sense of time: run backwards it in no way violates common sense as did the examples in the last chapter. The arrow of time discussed there had two important elements. In an isolated system, entropy reducing trajectories are exceedingly sensitive to initial conditions. In this respect they are similar to chaotic ones. The second important consideration in that discussion – that in equilibrium the overwhelming majority of trajectories are not entropy reducing – is not present in systems of a few particles. To put it another way, a chaotic trajectory aimed very slightly differently gives a quantitatively very different, but a nonetheless qualitatively similar, chaotic one. Aim an entropy reducing trajectory in ever so slightly a different way, however, and you get one that contributes to the increase of entropy.

In this introduction to a currently active research topic, I have strayed from my professed aim of doing specific examples. In the next two sections I attempt redemption by analyzing two simple mathematical models, each of which has a sensitive dependence on initial conditions. The second can also be turned into a random number generator, which – since it is obtained from a chaotic but fully deterministic prescription – should more correctly be called a pseudo-random number generator.

---

† Important contributors to these calculations are the contemporary astronomers Jack Wisdom and Jacques Laskar.
‡ For example, the impact of the large meteor on the earth sixtyfive million years ago which is thought to have finished off the dinosaurs and the well-observed collision of the comet Shoemaker-Levy 9 with Jupiter in July 1994.

A rudimentary pinball machine

Anyone older than video games will remember devices in which a small metal ball was projected or allowed to fall into an obstacle course of small pins, and eventually reached one of several holes. 'Rewards' were assigned in such a way that more valuable outcomes were less likely. The natural occurrence of the last word indicates that such games are an excellent illustration of how a sensitivity to precise initial conditions converts something deterministic into an arena for the workings of luck. Here we consider an idealization of such a game chance for which an instructive calculation is possible.

In a problem at the end of Chapter 6 we considered a mass M on a spring and showed that the total energy

$$E = \tfrac{1}{2}Mv^2 + \tfrac{1}{2}kx^2 \tag{11.1}$$

remained constant in the course of the motion. The first term ($\tfrac{1}{2}Mv^2$) is the kinetic energy of the mass M with velocity v, and the second ($\tfrac{1}{2}kx^2$) is the potential energy associated with stretching the spring by an amount x. [The constant k describes the stiffness of the spring.]

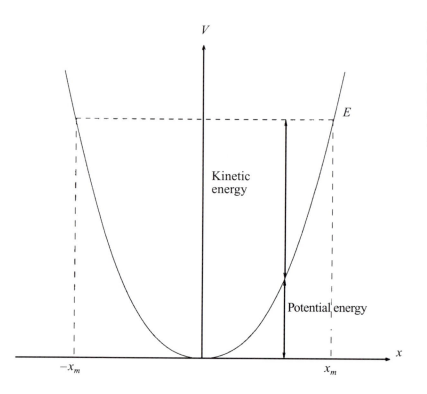

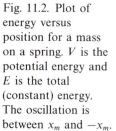

Fig. 11.2. Plot of energy versus position for a mass on a spring. V is the potential energy and E is the total (constant) energy. The oscillation is between x_m and $-x_m$.

We can visualize the motion by making a plot of the potential energy, a parabola in x, as in Fig. 11.2. The constant energy, E, is shown as a horizontal dashed line. This energy is shared between kinetic energy and potential energy as shown. Since both of these are quadratic forms and never negative, the particle moves back and forth between x_m and $-x_m$, at which points of maximum displacement the energy is purely potential. Suppose now that there is some friction in the system, gradually converting some of the energy into heat. Under these circumstances, E will slowly decrease, as will the maximum amplitude x_m, till eventually the particle comes to rest at $x = 0$.

Now consider motion in a potential with a small hill near $x = 0$ and two neighboring valleys. This is illustrated in Fig. 32, where the height of the hill is taken to be E_0, and the valleys are at $\pm x_0$. If friction can be ignored, an energy greater than E_0 means that the system can surmount the hill and move between positive and negative x. In contrast, an energy less than E_0 means that the system is necessarily trapped in one of the valleys. If we now include friction and start the

Fig. 11.3. Potential energy of the 'Pinball Machine'. E_0 is the energy needed to reach the top of the hill.

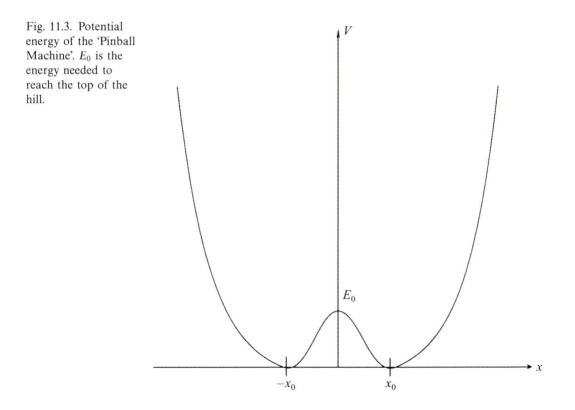

system off with an energy greater than E_0, which valley it ends up in depends sensitively on how it is started.

As you might guess, a knowledge of the initial energy is not sufficient to decide this question. Since friction reduces the energy, passages over the hill become progressively more laborious – by which I mean that the kinetic energy and thus the speed at each successive traversal of the peak become smaller – until finally trapping occurs in one of the valleys. In fact, the initial position and velocity (which determine the initial energy, but not *vice versa*) are needed to determine the motion. To do a calculation, one needs a model for the friction. The simplest assumption is to describe it by a force opposing the motion and proportional to the instantaneous speed. It is then possible to solve the problem numerically, making appropriate choices for the shape of the potential and the strength of the friction.

The results of such a calculation are shown in Fig. 11.4. A number of trajectories are shown in a plane in which position is plotted horizontally and velocity vertically. Before discussing the information in this curiously complicated figure, it is worth making sure that its meaning is clear. The coordinates of any point in the plane specify a position and a velocity. Points in the upper right quadrant correspond to being in the right half of Fig. 11.3 and moving towards the right. [We are taking a rightward velocity to be positive.] Similarly, a point

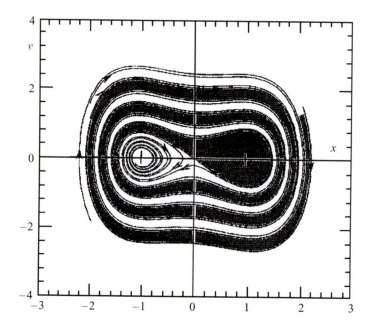

Fig. 11.4. Plot of position vs velocity for a particle subject to friction and moving in the potential of Fig. 11.3,

in the upper left quadrant, where x is negative and v positive, means that the particle is on the left but moving towards the right. And so on for the other two quadrants. A directed line in the plane gives the position–velocity history of a trajectory. Four different numerically calculated motions are shown in the figure.

Consider the trajectory shown starting in the upper right quadrant. This has been aimed in just such a way that after surmounting the hill three times from right to left (interleaved with three transits from left to right) it finally slowly climbs the hill from the right, just makes it, and comes to rest balanced at the very top with an energy of precisely E_0. The trajectory shown starting in the lower left is a similar very special one having the same precarious final destination but approaching it from the left.† All other trajectories eventually settle into one of the valleys. Two which both end in the left are shown. One (with an energy ever so slightly lower than the special trajectory which makes its last approach to the top from the left) just fails to clear the hill from the left and spirals down to the left. Another (with an energy ever so slightly more than the special trajectory which makes its last approach form the right) squeaks across the top ($x = 0$) from the right with a small negative (i.e. left moving) velocity and also eventually settles in the left valley. The complicated jelly roll shows how small initial differences are magnified by the motion. The shaded region in the figure includes all trajectories that end in the right valley. It has precisely the same area as the other half of the jelly roll which includes all trajectories that end in the left valley. An initial condition not chosen with special care thus has an equal chance of yielding a final destination in the left or the right, and probability thus emerges from imperfectly specified perfectly determined motion.

Logistics

The previous example illustrated how sensitive orbits can be to exact starting conditions, but it was also rather special and smooth. In general there is nothing analogous to the valleys that brought many diverging trajectories together again. Our rudimentary game of chance therefore did not illuminate an important element underlying the difficulty of predicting chaotic orbits, which is that errors accumulate

† You will protest that these delicately aimed trajectories are absurd: the slightest inaccuracy or jitter from any source will prevent them from stopping at the very top of the hill. To this I have two replies: (i) The model we are considering – like others considered previously, e.g. the ideal gas – is artificial but instructive; (ii) These special trajectories can be thought of as hypothetical watersheds which separate the two valleys.

multiplicatively (as in compound interest) and lead to an exponential increase in the inaccuracy of prediction. Here we shall consider a now famous model which displays these troublesome characteristics.

The model is called the 'logistic map,' which, as the name hints, has origins in the theory of populations. One is to imagine a population (of insects for example) which grows by reproduction but is limited by crowding. Let N be the population in one breeding cycle, and N' in the next. One argues that an increase is proportional to the present number, whereas a decrease (due to a shortage of food and living space) depends on interactions between individuals. Now, the number of pairwise interactions is $N(N-1)/2$ which is proportional to N^2 for large N. The model is thus summarized by the deceptively simple equation

$$N' = gN - dN^2. \tag{11.2}$$

Here g and d are two positive numbers describing the growth and decay of the population. The first is the more interesting parameter; d determines the maximum viable population, but has no other interesting effect. Note that if $N = g/d$, (11.2) leads to $N' = 0$. Measuring N as a fraction of this theoretical limit, i.e. defining $n = Nd/g$, with a similar definition of n' reduces (11.2) to

$$n' = gn(1 - n). \tag{11.3}$$

from its meaning one notes that the allowable range of n is $0 \le n \le 1$.

That the properties of this simple equation depend strikingly on the 'control parameter' g was shown most completely by Mitchell Feigenbaum (a physicist) in the late 1970s following earlier work by Robert M. May (a physicist turned mathematical ecologist), Stephen Smale (a mathematician), and others. They showed that increasing g leads to progressively less self evident connections between successive generations, culminating in chaotic behavior where wild fluctuations very effectively mask the simplicity of the underlying rule (11.3). Qualitatively and in a much simpler context, this is the phenomenon discovered by Poincaré: determinism in detail; apparent randomness to all but the best informed observers.

The onset of chaos here is called a 'period doubling cascade,' for reasons that will soon become clear. The intricate details would take me rather far afield from my subtitle: 'Probability and its uses in physics,' but a little investigation of the phenomenon is easy and instructive.

Let us ask if, for a fixed g, there are values of n which remain steady, i.e. for which $n' = n$. Solve (11.3) under this condition, and the answer is $n = 0$ or $n = n^* = 1 - (1/g)$. Since n must be greater than zero, the second solution only makes sense for $g \geq 1$. So we might guess for $g < 1$ that under the influence of the rule (11.3) the population steadily diminishes and reaches zero, and for $g > 1$ that the population tends towards n^* and reaches a steady state at this value. The first guess is true, the second is not, at least not in general. What we have failed to do is to analyze the stability of the second solution. How do small deviations from n^* propagate? Try $n = n^* + \Delta n$. Since $(n^* + \Delta n)^2 = (n^*)^2 + 2n^* \Delta n + ..$ we see to linear order in Δn that

$$\Delta n' = g\Delta n - 2gn^* \Delta n = -(g - 2)\Delta n, \tag{11.4}$$

where in the last form we have inserted the calculated expression for n^*. Thus, if n is slightly different from n^*, n' is on the other side of n^* but closer to it only if $(g - 2) < 1$. So something happens as g crosses 3. This is easily investigated using a small calculator. The surprising result is that instead of settling down to a steady state, a population emerges which oscillates back and forth between two values. Once one knows this result, it is quite easy to deduce by calculation. Instead of $n' = n$, we now want $n'' = n$ where n'' is related to n' by a reapplication of the same rule (11.3). The condition $n'' = n$ gives the equality

$$n = n'' = gn'(1 - n') = g[gn(1 - n)][1 - gn(1 - n)]. \tag{11.5}$$

The first and the last expressions give a rather formidable looking equation determining possible values of n which return to themselves after two breeding cycles. But a great simplification is possible because we know that the equation must have $n = 0$ and $n = n^*$ as solutions, because for these cases we already know that $n' = n$, from which it follows that $n'' = n' = n$. Guided by this knowledge, we can find the remaining solutions by solving a quadratic equation. I shall spare you the algebra and give the result. The two new solutions are

$$n_{1,2} = \frac{1}{2g}[(g + 1) \pm \sqrt{(g - 3)(g + 1)}] \tag{11.6}$$

Note that for $g < 3$ the square-root is of a negative number, and so cannot determine a sensible population (as a fraction of the maximum sustainable one). So the new solutions only make sense for $g > 3$. Note also that for $g = 3$, $n_1 = n_2 = \frac{2}{3}$, which is also the value of n^* for $g = 3$. Thus for g slightly greater than 3, the stable population n^* 'bifurcates' to yield a population alternating between two values in successive breeding cycles. Starting the system off with $n \neq n_1$ or n_2 leads quite quickly via the application of (11.3) to the just calculated 'stable 2-cycle.' [This is explored in a problem.]

As before, one must analyze the stability of this solution. We have to ask when $\Delta n''$ is larger in magnitude than a small deviation of n from n_1 or n_2. This calculation can be done and gives the result that the 2-cycle becomes unstable at $g = 1 + \sqrt{6} = 3.44949\ldots$ What happens then? Now the algebra starts to get hopeless, but a calculator quickly shows that a 4-cycle ensues. As g is further increased, an 8-cycle emerges. This pattern (period doubling) goes on for ever, but the g values for successive period doublings occur faster and faster. One finds that if g_n is the value at which there is a transition from 2^{n-1}-cycle to a 2^n-cycle, $\lim(n \to \infty)g_n \approx 3.57$. More interestingly, Feigenbaum discovered that the difference between gs for successive bifurcations approaches a constant ratio.

$$\frac{g_n - g_{n-1}}{g_{n+1} - g_n} = \delta, \tag{11.7}$$

where δ, now called the Feigenbaum number, is found by numerical methods to be $4.6692\ldots$ He also showed, intriguingly, that this number occurs not just in the logistic map but also in any transformation from n to n' which has a qualitatively similar (one quadratic-humped) structure.

Chaos by stretching and folding

In recent years, period doubling cascades have been found in electronic circuitry, in the flow of liquids, and in other situations having an externally controllable parameter analogous to the growth factor g of the last section. The universal properties of such cascades are extremely interesting. More central to our purposes, however, is what happens when g is increased beyond g_∞. Many things, as you might imagine. I refer you to the readings suggested at the end of this chapter for a wider view of this richly complicated region. Here, I want to consider the special point $g = 4$ for which relatively simple but cogent remarks are possible.

To explain why this point is special, I need a little trigonometry. In Fig. 11.5(a) is shown the geometrical meaning of the sine and the cosine. A right triangle with a hypotenuse of unit length has (by definition) sides of length the sine and the cosine of one of the subtending angles, the conventions being illustrated in the figure. The sine and the cosine are thus numbers that depend on (i.e. are functions of) the angle θ; they are conventionally written $\sin\theta$ and $\cos\theta$. The

Pythagorean theorem evidently implies $\sin^2\theta + \cos^2\theta = 1$ for all θ. [The square of a trigonometric function is conventionally indicated by putting the superscript 2 where I have it.] Now consider the drawing in Fig. 11.5(b). It shows an isosceles triangle obtained by putting together the triangle of Fig. 11.5(a) and its mirror image. I wish you to recall that the area of a triangle is $\frac{1}{2}\times$ (base) $\times$ (perpendicular height). Evidently the triangles OAB and OBC each have an area of $\frac{1}{2}\sin\theta\,\cos\theta$. So the area of OAC is $\sin\theta\,\cos\theta$. But there is another way of calculating this area. AD is a perpendicular dropped from A to the side OC. Its length is $\sin(2\theta)$. [Why?] Further, OC has unit length by construction. Thus the area OAC is $\frac{1}{2}\sin(2\theta)$. Collecting pieces, we thus have

$$\sin(2\theta) = 2\,\sin\theta\,\cos\theta, \qquad\qquad (11.8)$$

which is all the trigonometry we need.

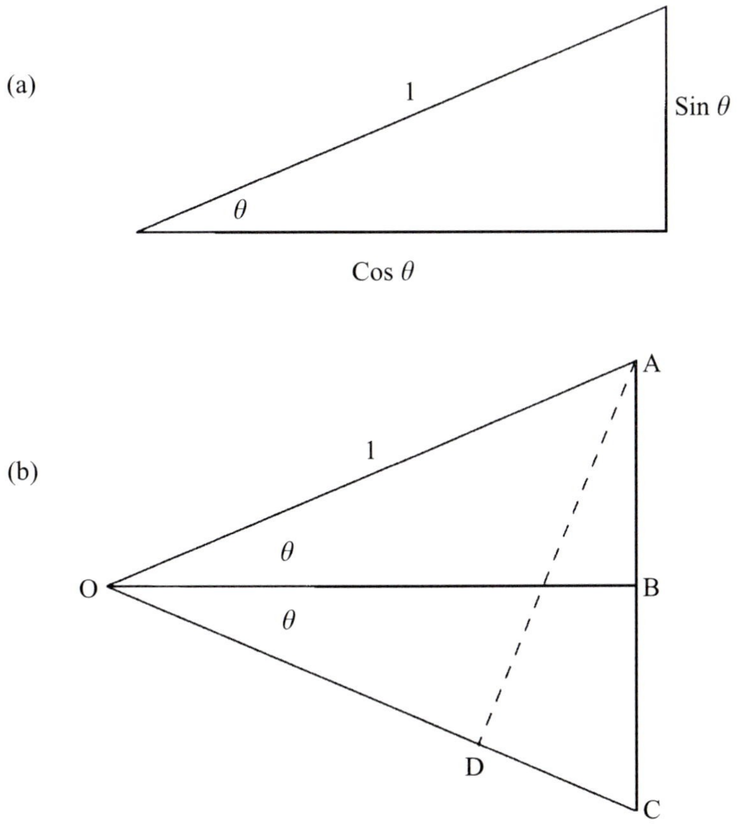

Fig. 11.5. (a) a right triangle illustrating the definition of $\sin\theta$ and $\cos\theta$ (b) construction needed to prove (11.8).

Return to the logistic map for $g = 4$, namely

$$n' = 4n(1 - n). \tag{11.9}$$

This is plotted in Figure 11.6. Note that as n goes from 0 to 1, n' traverses this region twice. The regions $0 \leq n \leq \frac{1}{2}$ and $\frac{1}{2} \leq n \leq 1$ are both stretched to $0 \leq n' \leq 1$ and the two are folded together, rather as a baker might do in making a flaky pastry. One might guess that this operation repeated over and over would scramble the numbers from 0 to 1 rather well. It does, in a way we are about to understand.

Now comes a mathematical trick, i.e. a useful transformation guided by intuition rather than deduction. Introduce an angle θ such that $n = \sin^2 \theta$. This is possible because n, $\sin \theta$, and $\sin^2 \theta$ all run between 0 and 1. Substitute into (11.9) to get

$$\begin{aligned} n' &= 4 \sin^2 \theta (1 - \sin^2 \theta) = 4 \sin^2 \theta \, \cos^2 \theta \\ &= [2 \sin \theta \, \cos \theta]^2 = \sin^2(2\theta). \end{aligned} \tag{11.10}$$

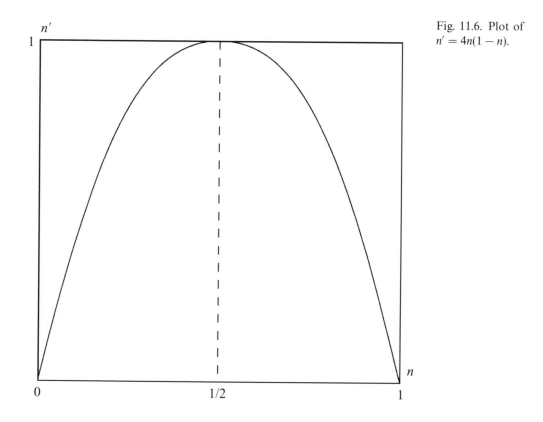

Fig. 11.6. Plot of $n' = 4n(1 - n)$.

Since n' is connected to 2θ exactly as n is connected to θ, we see that underlying the stretching and folding operation of the logistic map for $g = 4$ is the repeated doubling of an angle. Now, increasing an angle corresponds to rotating the hypotenuse of a right triangle. (See Fig. 11.5(a)). Every complete rotation brings the hypotenuse and thus the sine and the cosine back on themselves. Whenever doubling an angle makes it bigger than 2π it is thus natural to subtract 2π so that the angles always lie between 0 and 2π. Equivalently, we may write $\theta = \alpha/2\pi$ where α is a number between 0 and 1.

To summarize, by writing

$$n = \sin^2(\frac{\alpha}{2\pi}) \tag{11.11}$$

we have been able to represent repeated applications of the rule (11.9) as

$$\alpha \rightarrow 2\alpha \rightarrow 2^2\alpha \rightarrow \ldots 2^n\alpha \rightarrow \ldots, \tag{11.12}$$

where whenever one member of the this sequence becomes larger than 1, the 1 is to be subtracted.[†]

Now we are in a position to show that how n or equivalently α develops depends sensitively on the starting point. This is most easily seen by writing α as a binary number, i.e., as

$$\alpha = \frac{p_1}{2} + \frac{p_2}{2^2} + \frac{p_3}{2^3} + \ldots \tag{11.13}$$

where each of the coefficients, p_1, p_2, etc. is 0 or 1. In analogy to decimal notation, we may write the number α as $0.p_1 p_2 p_3 \ldots$. Any number between 0 and 1 can be expressed this way. For example, $\frac{1}{2}$ is 0.1 and $\frac{1}{16}$ is 0.0001. These reciprocals of multiples of 2 have very simple binary representations, just as reciprocals of multiples of 10 are simple decimals. Other binary representations are $\frac{1}{3}$ as 0.010101..., and $\frac{1}{5}$ as 0.00110011..., where the dots mean an endless repetition of the displayed pattern.

The advantage of the binary representation for our purposes is that multiplication by 2, as you will readily verify, moves the point one position to the right. So, successively moving the point one step to the right gives the transformation of α under the $g = 4$ logistic map. [Note that for the reason given above, a 1 appearing to the left of the point is to be dropped.] Thus the sequence (11.12) corresponds to extruding and chopping the number, rather like a left handed cook

---

[†] Note that a small error in α will be magnified by a factor $2^n = e^{n \ln 2}$ after n iterations. This is the exponential loss of accuracy mentioned earlier.

chopping an endless carrot. Evidently the process quickly probes the deep structure of the number on smaller and smaller scales.

Any computer stores numbers as a finite sequence of 0s and 1s. You might therefore think that on a calculator the rule (11.9) will eventually produce α and thus n equal to 0.00..., where it would sit forever. However, $n = 0$ is an unstable point, and the calculator working with finite numbers makes approximations in working out (11.11) and (11.9). So, entering an arbitrary number in a calculator programmed to iterate (11.9) gives an apparently random sequence. This is shown in solved problem 4, which also demonstrates that very close starting points rather quickly wander away from each other.

To discuss the generation of random numbers in principle by this process one need only contemplate irrational numbers, which when expressed in binary form have no end and no repeating pattern. [See solved problem 2.] Give me an irrational number between 0 and 1 (such as $\pi - 3$) and I can work out its binary representation, from which I can deduce exactly how this 'seed' will determine the workings of the $g = 4$ logistic map. Show me the output as a sequence of ns and I would be very hard put to deduce the underlying simple rule. This is (deterministic) chaos very concretely displayed.

A brief look back and forward

With chaos, this book finally but firmly enters the last quarter of the twentieth century. It is an interesting episode in the history of science that this phenomenon, lurking around as a worry for theoretical and practical astronomy for more than a century, has made its presence felt more or less simultaneously in many fields. The widespread availability of more and more powerful computers is responsible for the burgeoning interest. This has also led some to express the view that chaos hammers the last nail into the coffin of eighteenth century determinism. It is true that Nature as an eighteenth century piece of simple and regular clockwork has been ailing for some time. But there is no need to exaggerate. Astronomical orbits can be calculated with enough precision for every practical purpose, and indeed accurately enough to *predict* chaotic behavior where it occurs. In the mechanics of large objects, chaos is a manifestation of Newton's Laws at work, even if sometimes, as with the weather, it very effectively masks any simple underlying principles. In the mechanics of the atomic and sub-atomic world the situation is rather different: regular and orderly motions when observed require probability for their interpretation. This last and perhaps most bizarre occurrence of probability in physics is the subject of the next and final chapter.

Further reading

There are a number of recent popular books and articles on Chaos. Here is a list of sources I have had scattered around my desk. It makes no pretence of completeness.

Perhaps the most dazzling popular book is *Does God Play Dice?* by Ian Stewart (Basil Blackwell 1989). Two other small volumes (from a rather more mathematical perspective than mine) are *Mathematics and the Unexpected*, by Ivar Ekeland (University of Chicago Press, 1988) and *Chance and Chaos*, by David Ruelle (Princeton, 1991). *Chaos: A View of Complexity in Physical Science*, by Leo P. Kadanoff (short and to the point) is to be found in the 1986 supplement [Great Ideas Today 1986] to the *Encyclopædia Britannica*. The 1994 edition of this Encyclopædia has an article called Physical Science, Principles of, which ends with two wonderfully clear and authoritative sections called Entropy and Disorder, and Chaos under the initials ABP (Sir Brian Pippard, past Cavendish Professor in the University of Cambridge.)

Solved problems

(1) Show that $\sqrt{2}$ is not a rational number.

Solution: Suppose the contrary. Then one could write $\sqrt{2} = p/q$ where p and q are integers. Assume that any common factors (except of course 1) have been canceled out. Now square both sides to get the equation

$$2 = \frac{p^2}{q^2}, \text{ or } p^2 = 2q^2, \tag{11.14}$$

which shows that p^2 is twice the integer q^2 and thus divisible by 2, or even. Now p^2 being the square of an integer can only be even if p is even. So $p = 2s$ where s is an integer, and $p^2 = 2 \times 2 \times s^2$. It follows that $q^2 = 2s^2$, whereupon, using the previous argument, q is seen to be even. But p and q cannot both be even, because we arranged for them to have no common factors. The contradiction means that $\sqrt{2}$ cannot be written as a ratio of integers.

(2) Show that an irrational number cannot have a finite or repeating binary representation.

Solution: A number whose binary representation ends with an infinite string of zeros is clearly rational. For, if the last 1 occurs in the nth position, the number can be put on the common denominator 2^n. A number with a pattern that exactly repeats over and over again is also rational. Consider the number 0.00110011.... The first 4 digits can be put on the common denominator $2^4 = 16$. In fact $0.0011 = \frac{3}{16}$. But now the next 4 digits give the same result multiplied by $(1/2^4)$. Thus the infinitely repeated pattern corresponds to an infinite series, which can be summed. In detail

$$\frac{3}{16}\left[1 + \frac{1}{16} + (\frac{1}{16})^2 + \ldots\right] = \frac{3}{16}\frac{1}{1 - \frac{1}{16}} = \frac{3}{15} = \frac{1}{5}. \qquad (11.15)$$

A similar evaluation is possible for any repeated pattern, always leading to a rational fraction. Thus no irrational number has a finite or repeating binary representation. [Actually the proof shows that the result is true regardless of the base of the number representation.]

(3) On a hand calculator, try out the logistic map for various values of the growth parameter in the period doubling region. Guided by the analysis in the text, I suggest $g = 0.5$, 2.0, 3.1, and 3.5.

Solution: With $g = 0.5$, and starting with $n = 0.5$, I found a steady decrease to $n = 0.000$ in about 10 iterations. Since g is less than 1, the stationary point is as expected 0.

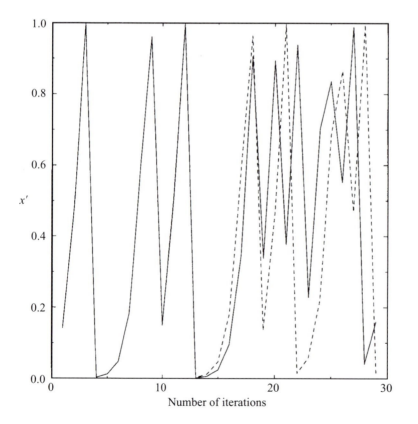

Fig. 11.7. Iteration of $x' = 4(1 - x)$ with different initial values of x: for the solid line $x_0 = 0.14159$ whereas for the dashed line $x_0 = \pi - 3$.

Number of iterations

With $g = 2$, and starting at 0.9, I got the sequence 0.1800, 0.2952, 0.4161, 0.4859, 0.4996, 0.5000, and no change thereafter. This is the fixed point $1 - (1/g)$ expected for $g < 3$.

With $g = 3.1$ and starting at 0.8, I had to go through about 20 iterations before settling into a 2-cycle at 0.7646 and 0.5580. These are exactly the numbers given by (11.6) to 4 figures.

With $g = 3.5$, the iterations settled into a 4-cycle with the numbers 0.5009, 0.8750, 0.3828, and 0.8269 repeating in the same order.

(4) Consider the case of $g = 4$ first entering the number $\pi - 3$ into the calculator, but requiring it to show only 4 figures, and then starting from 0.1416, explicitly entered in this form.

Solution: What is interesting about this game is that the calculator shows the very same starting number in both cases. However, there are digits not shown but carried in the memory in the first case (where the number entered is actually 0.141592654). The results are plotted in Fig. 11.7. The two sequences look very similar for about 10 iterations but then start to bear no relations to each other, a manifestation of extreme sensitivity to initial conditions.

12

Quantum jumps: the ultimate gamble

> Even the experts do not understand it the way they would
> like to, and it is perfectly reasonable that they should not,
> because all of direct, human experience and of human
> intuition applies to large objects.
>
> Richard P. Feynman

Two great intellectual triumphs occurred in the first quarter of the twentieth century: the invention of the theory of relativity, and the rather more labored arrival of quantum mechanics. The former largely embodies the ideas, and almost exclusively the vision, of Albert Einstein; the latter benefited from many creative minds, and from a constant interplay between theory and experiment. Probability plays no direct role in relativity, which is thus outside the purview and purpose of this book. However, Einstein *will* make an appearance in the following pages, in what some would claim was his most revolutionary role – as diviner of the logical and physical consequences of every idea that was put forward in the formative stages of quantum mechanics. It is one of the ironies of the history of science that Einstein, who contributed so importantly to the task of creating this subject, never accepted that the job was finished when a majority of physicists did. As a consequence, he participated only indirectly† in the great adventure of the second quarter of the century: the use of quantum mechanics to explain the structure of atoms, nuclei, and matter.

Here in outline are the contents of this final chapter. My general aim is to give an honest glimpse – I shall try not to do less; under the circumstances, I do not know how to do more – into what quantum mechanics explains, how it uses the idea of probability, how it fits into statistical mechanics, and ultimately, how *odd* it is. First, I discuss some properties of waves and describe experiments which show light behaving like a wave. Then I talk about the 'photoelectric effect,' in which light behaves like a particle. That light is in fact *both* emerges from a discussion of photography under very weak illumination. Here probability emerges as the best quantum mechanics has to offer for

---

† The *important* exception is Bose–Einstein condensation, a topic which is outside the scope of this chapter

where a quantum event occurs. Finally, I discuss how quantum and statistical mechanics are put together, briefly discuss how a laser works, and indicate that *when* a quantum is emitted is also answered only in a probabilistic way. The picture that emerges deeply troubled Einstein, and troubles some thoughtful people to this day. However, no one has found a fundamentally less paradoxical theory which is not in conflict with experimental facts. And it is with probability, not as the consequence of missing information but as the best prediction of what is generally believed to be a complete description of Nature, that this tour of the uses of chance in physics comes to a close.

Even more than in previous chapters, I am here trying to do too much, much too quickly. An even remotely systematic treatment would require considerably more space. My hope, to take refuge in a thought expressed in the preface, is to slightly open a door that might otherwise remain closed. Other readings are suggested at the end of the chapter.

Classical particles and waves

Quantum mechanics was invented to explain phenomena involving the interplay of light and electrons. Electrons were discovered by the English physicist J. J. Thomson (1856–1940) in 1897. They are the most common and most easily moved electrified elementary particles in Nature. Either electrons or light can be used to illustrate quantum mechanics, and both behave equally strangely when it is at work. In what follows, I have chosen to give light the primary role because, our eyes being light detectors, it is familiar. Although special equipment is needed to detect them, electrons are ubiquitously at work in the modern world. Light illuminates the book you are reading; electrons carry the electrical current to your reading lamp. Light emanates from the screen of a television set; electrons carry the message within a television tube.

In the pre-quantum world light was purely a wave and electrons simply particles. Quantum mechanics tells us that light can also be particle-like and electrons can also be wave-like. Both, as Richard Feynman (1918–1988) – perhaps the most inventive physicist of his time – put it, are 'wavicles.' Before one can grasp this idea, or explore its further strangenesses, it is necessary to understand what particles and waves meant in 'classical,' i.e. nineteenth and early twentieth century, physics. In this era the distinction between the two concepts was very clear.

Classical particles have been repeatedly encountered in this book.

They obey Newton's equations (Chapter 6.) A particle is characterized by where it is and how it moves, i.e. by a position and a velocity. If it encounters a wall with two holes, a particle might go through one hole or the other. The notion that it might go through both simultaneously without breaking up is, from a Newtonian point of view, absurd.

As for classical waves, they can be understood by considering ripples on the surface of water. Such ripples, or waves, spread out in all directions: it is in the nature of a wave that it is not necessarily confined at any given time to a small region of space as is a classical particle. A snapshot of a simple wave, called a 'plane wave' is illustrated on the left of Fig. 12.1. Crests are indicated by solid lines and hollows by dashed lines. As time progresses the entire pattern is to be imagined moving steadily to the right. A plane wave can be characterized by a 'wavelength' (crest to crest distance), an 'amplitude' (half the crest to trough height), and a 'wave velocity.' To understand what is meant by the velocity of a plane wave, let λ be the wavelength, and τ the time (called the 'period') it takes for a particular crest to occupy the position of the previous crest at time 0. The magnitude of the wave velocity is then given by $c = \lambda/\tau$, and its direction is the direction of motion of the crests. It is common to characterize waves by their 'frequency,' defined as the reciprocal of the period, and often called v. Evidently,

$$c = \lambda v. \tag{12.1}$$

In general, how a wave propagates depends on its amplitude. We shall assume here that waves of all amplitude behave in the same way. A system for which this assumption is correct is said to obey the principle of superposition, and is called 'linear.' We shall make this assumption in what follows.

If a wave encounters a barrier with two apertures, it *can* go through both of them at the same time. On the other side of such a barrier one observes a phenomenon called 'interference.' Imagine a plane wave whose crests and troughs are parallel to the barrier impinging on it from the left as illustrated in Fig. 12.1. [Such a continuous train of waves would be set up not by just dropping one stone into a still pond but by forcing the surface to oscillate up and down far to the left of the figure.] To the right of the barrier the wave is no longer plane. Instead, the crests and hollows are concentric circles emanating from the apertures, because it is only in these regions that the water on the right is being forced up and down. How do such separated wave trains recombine? The important point, explained in detail in the next paragraph, is that depending on the direction of observation the wave-crests from the two holes are in step, partially out of step, or totally out of step (so that the crests from one side match up with the hollows from the other). As a consequence, a row of buoys parallel to but far to the right of the barrier will register periodic maxima and minima – called an interference pattern – in the height of the wave oscillations.

Interference allows a quantitative determination of the wavelength of a wave. This is best understood by calculating the interference pattern for the barrier with two apertures. The construction shown in Fig. 12.1 determines one direction in which the two emerging wave trains 'interfere constructively,' i.e. are in step. As shown, θ is the angle between this direction and a perpendicular to the barrier. Now, consider the right triangle ABC. It follows from elementary geometry that the angle ABC is also θ. The side AC is evidently the wavelength, which we shall call λ. The separation between the openings AB will be called d. By the definition (Fig. 11.5) of the sine of an angle, we see that $\sin \theta = (\lambda/d)$. This is evidently the angle for which a wave crest from the upper opening matches up with the previous crest from the lower one. Constructive interference also occurs when the separate wave-trains differ by any whole number of wavelengths. The general

Fig. 12.1. Illustration of a two-slit interference pattern.

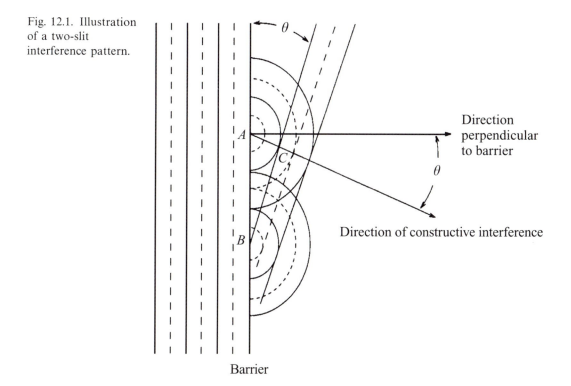

condition for constructive interference is thus

$$\sin \theta = \frac{n\lambda}{d},\qquad(12.2)$$

with n being a positive or negative integer or zero.

The maximum amplitudes in the interference pattern are thus predicted to occur at angles given by (12.2), both sides of which are plotted in Fig. 12.2 for θ between 0 and 90°. If d is known and the pattern observed, this equation allows λ to be determined. Note the interesting possibility $n = 0$, corresponding to $\theta = 0$, which says that in the direction directly forward of the barrier the waves are in step.

Light as a wave

Very interestingly, an interference pattern *is* observed when light illuminates a screen after passing through two slits (of suitable dimensions) in an opaque barrier. [This was first shown by the English scientist Thomas Young (1773–1829).] The image on the screen is a series of separated single sharp lines determined by (12.2), providing strong evidence that light is a wave.† In fact, the wavelength of light was

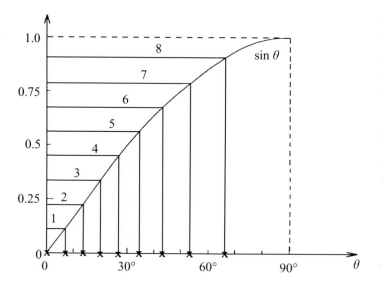

Fig. 12.2. Graphical construction for angles of constructive interference. The ratio λ/d has been chosen to be a little over $\frac{1}{9}$; the labels on the horizontal lines label the n of (12.2) and the crosses on the vertical lines the corresponding angles.

† From the discussion of water waves it is evident that for two-slit interference to be observed with light the spacing between the openings must be comparable to the corresponding wavelength. It is also necessary that the width of each opening be small (to minimize 'diffraction,' the word for interference between different parts of the wave in one opening) and for the light to be in the form of a long wave train – most easily achieved in the modern world by using a laser.

first measured by observing interference. To see the effect clearly, one uses light of a definite color, because what we perceive as colors are in fact different wavelengths. A deep red is found to have a wavelength of about 0.7 μm (micrometers).‡ An intense violet has a shorter wavelength of about 0.4 μm. In between is the rainbow of colors.

It is worth mentioning that the properties of light traveling in anisotropic media – transparent materials in which not all directions are equivalent – were completely explained by taking light vibrations to be perpendicular to the direction of wave motion: a three dimensional generalization of the transverse motion of water in surface ripples. The explanation of light propagation in media and of interference had by fairly early in the nineteenth century made it unreasonable to think of light as anything but a wave, though what exactly was vibrating remained a mystery.

The final triumph of the wave theory was the prediction in 1862 by James Clerk Maxwell (1831–79), as a consequence of linear equations formulated by him, that 'electromagnetic radiation' of all wavelengths could propagate in what he called the æther and we now call – without to this day fully understanding the medium that sustains these vibrations – the vacuum. In Maxwell's waves electric and magnetic 'fields,' which exert forces on electrically charged objects, oscillate perpendicularly to the direction of transmission. From measured electrical and magnetic forces he estimated the speed at which these waves would travel to be very close to the speed of light – reasonably well known at the time from astronomical observations and terrestrial experiments. He hypothesized that ' ... light consists in the transverse undulations of the same medium, which is the cause of electric and magnetic phenomena.' Maxwell's equations also describe the generation and the interaction with matter of classical electromagnetic waves, thereby allowing their use and control.

In 1888, some 25 years after their prediction, Heinrich Hertz (1857–94) produced and detected electromagnetic waves using apparatus exploiting electrical sparks. We now know that visible light corresponds to only a very small wavelength region of the 'electromagnetic spectrum.' Going from the shortest to the longest wavelengths, γ-rays, X-rays, ultra-violet radiation, visible light, infra-red radiation, microwaves, and radio waves are electromagnetic waves. All have a

---

‡ 1 μm $= 10^{-6}$ m. [See the dictionary at the end of chapter VI.] For purposes of comparison: a fine human hair has a width of about 25 μm.

velocity equal to that of light (usually called c and equal to approximately 3×10^8 meters/second).†

The bright spot

Constructive interference in the forward direction, of which a particular case is the $n = 0$ line in the two-slit interference pattern, underlies the remark on p. 2 that the wave theory of light predicts a bright spot in the center of the shadow cast by a circular screen. The history of the discovery of this quintessential property of waves is most unusual, and tempts me into a digression which I shall use in what follows to illustrate quantum mechanics.

As background, it is necessary to know that in the eighteenth century, before the event to be described here and long before Maxwell's decisive theory, opinion was divided between light being a particle – or 'corpuscle' – and light being a wave. The corpuscular view owed much to Isaac Newton and his book *Opticks* (1704). Newton is justly remembered for his experiments on light and its colors. By his own account, he was more interested in phenomena than hypotheses, but apart from one somewhat ambiguous remark he thought of rays of light as a shower of particles. Perhaps he could not imagine that light could have so tiny a wavelength. Indeed, he wrote in 1675: 'If it consisted in pression or motion, propagated either in an instant or in time, it would bend into the shadow.' This remark suggests that he would have understood that interference and diffraction show light doing exactly that. In any event, Young's ideas were severely attacked by Newton's disciples.

Among the proponents of the wave theory in the early nineteenth century, besides Young, was Augustin Fresnel (1788–1827), who in 1818 submitted a memoir on diffraction for a prize offered by the French Academy. Sir Edmund Whittaker in his *A History of the Theories of Aether and Electricity* describes the reception of the memoir as follows:

> ... Laplace, Poisson and Biot, [eminent scientific thinkers of
> the time] who constituted a majority of the Commission to which
> it was referred, were all zealous supporters of the corpuscular

† To honor Hertz, 1 reciprocal second is called 1 hertz (written Hz). 1 MHz means one megahertz which equals 10^6 Hz. Thus a 100 MHz FM radio station is transmitting its message by F(requency) M(odulating) – or dithering the frequency of – an electromagnetic wave of wavelength [via (12.1)] $c/\nu \approx 3 \times 10^8$ (m/s) $/100 \times 10^6$ Hz = 3 m. By contrast, a typical medical X-ray has a wavelength of a fraction of 1 nm = 10^{-9} m.

theory. During the examination, however, Fresnel was vindicated in a somewhat curious way... Poisson, when reading the manuscript, happened to notice that the analysis could be extended to other cases, and in particular that it would indicate the existence of a bright spot at the center of the shadow of a circular screen. He suggested to Fresnel that this and some further consequences should be tested experimentally; this was done and the results were found to confirm the new theory. The concordance of observation and calculation was so admirable in all cases where a comparison was possible that the prize was awarded to Fresnel without further hesitation.

The experiments were done by Dominique-François Arago (1786–1853). I mentioned this episode in Chapter I as an example of how science progresses; it is also a paradigm for how scientific argument ought to be, but is not always, conducted. To be sure, the issues are seldom so clearly posed.

Light as a particle

During experiments demonstrating the existence of electromagnetic waves, Hertz detected a phenomenon which we now know as the photoelectric effect. It consists in the ejection of electrons from metal surfaces bathed with light.

In 1905, Einstein provided a quantitative explanation of the effect which for the first time posited that light comes in quanta of energy, now called photons.† In a deeply original way, Einstein attributed packets of energy $h\nu$ to light of frequency ν, i.e.

$$E(\nu) = h\nu. \tag{12.3}$$

Here h is a constant with the dimensions of energy multiplied by time which had been introduced into physics by Max Planck in 1901 in a context to be discussed later in this chapter. It is now known to be approximately equal to 6.63×10^{-34} joule/hertz, a number the extreme smallness of which explains why quantum effects are not easily noticed. Based on his photon hypothesis, Einstein predicted that the maximum energy of electrons ejected from a given surface by the action of light is given by

$$E_{max} = h\nu - W. \tag{12.4}$$

---

† It should be emphasized that this is not a re-emergence of the eighteenth century 'corpuscle.' This will become clear in the next section.

The picture is that an energy hv is transferred to an electron which then uses up at least W (which depends on the surface) to escape from the metal.

Equation (12.4) was completely confirmed by experiment by 1915. It is found to be obeyed *no matter how weak the illumination*; less light leads to fewer but not less energetic ejected electrons. This behavior is incomprehensible in a classical model of light, in which dim illumination means small electromagnetic vibrations and a weaker effect on electrons.

Planck started the quantum revolution in 1900. After a confusing transitional period of about 25 years, modern quantum mechanics was invented in a great burst of inspiration between 1925 and 1928, primarily by Heisenberg, Schrödinger, and P. A. M. Dirac (1902–84). In the quantum theory of electromagnetic radiation an oscillation of frequency v can only take or give up energy in Einstein's quanta of hv. This implies that there are tiny but discrete kicks associated with the absorption of such a wave no matter how weak it is. (Although a classical wave does exert a pressure on a wall that absorbs it, the amplitude of a classical wave can be reduced to zero and then the associated 'radiation pressure' also goes to zero.) In short, waves in quantum mechanics have a particle-like aspect.

This blurring of the distinction between waves and particles is not the truly puzzling thing about quantum mechanics, which in fact unifies the two concepts. It is in ascribing probabilities to processes in which no information is missing that quantum mechanics differs from every other physical theory we have.

Probability in quantum mechanics

The essential strangeness of quantum mechanics can be brought out by doing a thought experiment involving photography. The image you see when looking through the lens of your camera is recorded on the film by a quantum-mechanical process. The film contains a very large number of grains of a light-sensitive silver compound embedded in a gelatinous base. Photons can be absorbed in each grain. Every such process transfers a quantum of energy from the light to the material and causes the migration of an atom of silver to preexisting specks of metallic silver in the grain. After several such transfers to a particular speck, enough silver is accumulated to make the region react preferentially with the chemicals in the developer.

This description is confirmed by illuminating an object rather faintly and photographing it with a very short exposure time. The resulting picture is found to consist of a few dots, distributed apparently at random. One would say that only a few quanta of light – in other words, only a few photons – have entered the camera, so that only a few of the transitions described in the last paragraph have taken place.

Where on the film have the transitions occurred? Is there any message in the random dots? Study reveals something very interesting.† Imagine collecting thousands of such severely underexposed films of the same object taken with the same camera in the same position. Each of the films when developed shows a few dots. Construct a composite picture by marking the positions of all the dots from all the films on a single sheet of paper. One finds that the composite is a recognizable image of the photographed object! Evidently probability is at work here: an underexposed film is more likely to have developed specks in places where it is more intensely illuminated in normal exposure.

Is there a simple explanation for this phenomenon? One's first reaction is to assume that light simply *is* a swarm of classical particles, and argue that the properties of dim light follow from a proportionality between the number of these particles per unit volume and the intensity of the light in that volume: when only a few particles are admitted into the camera they would be more likely to be in the places where the light is intense.

This assumption is negated by photography under even weaker illumination. Planck's constant is so small that a normal light bulb radiates an enormous number of photons per second. [See solved problem 1.] On the other hand, the time it takes for a photon to be absorbed is rather short, so that it is possible to contemplate a situation in which only one photon enters the camera in the time it takes for one to be absorbed by the film. Experiments of this nature have been performed. With suitable very long exposures they yield normal photographs. Now, imagine a photograph made in this way of an interference pattern such as the bright spot at the center of a shadow cast by a circular screen. Since the bright spot can only be understood as constructive interference in the forward direction of a wave which simultaneously goes around all sides of the screen, and since it is very difficult to imagine how photons arriving one at a time could behave collectively, one is forced to the conclusion that the interference pattern is a property of a *single* photon.

---

† I have not done the experiment described here; nor can I give you a reference to a place where precisely this experiment has been reported. However, many closely related experiments have been done, leaving no doubt about the outcome of this one.

At this point, matters start to get mysterious. It is important to distinguish between the phenomena, which are strange, and the quantum mechanical explanation, which is stranger.

What happens if a very large number of underexposed photographs of a shadow containing a bright spot are made, each of which only allows a few hundred photons into the camera one at a time? Most of these photographs do not show the bright spot; some do. A composite made of the information in all of them is similar to a photograph made in ordinary illumination. More generally, any interference pattern can be recorded photon by photon. If the interference pattern is destroyed in ordinary illumination (for example by closing one slit of a two-slit arrangement) single photons also do not show it. These are the facts to be explained.†

Quantum mechanics comes with mathematical rules which have no difficulty with these facts. It says that in moving through space a photon is a wave, and like all waves can go around all sides of an obstacle at once. However, when it reaches the photographic plate it is registered in only one grain of the emulsion, i.e. it is a particle. To the question of *which* grain absorbs the photon, quantum mechanics gives a probabilistic answer. Regions that are intensely illuminated in ordinary light have a large probability of registering a single photon; correspondingly, regions that are totally unilluminated in ordinary light, for example regions where crests and hollows from different paths cancel each other out, have no probability of registering a photon.

The connection between the intensity of a classical wave and probability in quantum mechanics is worth dwelling on. The equations of quantum mechanics allow one to calculate 'amplitudes' for the passage of a photon from one point in space to another. For example, if L is a point to the left of a two-slit arrangement like Fig. 12.1 and R is a point to the right, one could in principle calculate an amplitude which we shall call $\Psi(L \to R)$ (Greek 'psi'). This quantity behaves like a classical wave amplitude. The amplitude for the photon to pass through the upper hole A is the product of the amplitudes for $L \to A$ and the amplitude for $A \to R$. A similar expression can be written for the amplitude corresponding to reaching R through the hole B. The principle of superposition holds in quantum mechanics: the total

---

† If this situation does not strike you as paradoxical, try to construct an explanation of why some dark parts of a two-slit pattern become illuminated when one of the slits is closed. The probability of a single photon hitting certain parts of the screen is actually *increased* by blocking one of the possible paths!

amplitude for $L \to R$ is the sum

$$\Psi(L \to R) = \Psi(L \to A)\Psi(A \to R) + \Psi(L \to B)\Psi(B \to R). \quad (12.5)$$

Now comes the strange part: the square of the amplitude is interpreted as the probability of the photon going from L to R. This has the immediate consequence that the probability of reaching R from L is *not* the sum of the probabilities of transits via A and via B, because there is a cross term in the square of the sum. Let P be the square of Ψ. Then, from (12.5) we have for this probability the expression

$$P(L \to R) = P(L \to A)P(A \to R) + P(L \to B)P(B \to R) \\ + 2[\Psi(L \to A)\Psi(L \to B)][\Psi(A \to R)\Psi(B \to R)]. \quad (12.6)$$

The interference of single quanta is built into the formalism, because the last product changes as the the wave fronts of the separate paths go from being in to being out of step. It is also a consequence of (12.6) that the interference disappears if one makes sure that the photon goes through hole A, by for example closing hole B so that the second term of (12.5) is not present. In short, forcing the photon to be particle-like – to go through one hole *or* the other but not both – destroys its wave-like property of interference, which is consistent with observations.

You will note that quantum mechanics explains the interference and probabilistic detection of individual photons without assuming any difference whatsoever between successively arriving ones. This is to be contrasted with every previous occurrence of probability in this book. A coin ends up heads or tails with equal likelihood because of an extreme sensitivity to initial conditions, which could in principle have been controlled. Here, however, no such crutch is available: probability is intrinsic to the basic theory.

It is fair to say that there is no generally accepted view of how to understand this matter. Referring to the photographic process one may ask when the transition from an amplitude like (12.5) to a probability like (12.6) occurs. Presumably it has happened once a particular silver atom has migrated to a speck. But quantum mechanics describes that process as well, so there is no point at which some bridge from amplitudes to probabilities is crossed ... This dilemma makes it tempting to imagine a more detailed theory in which the destination of a particular photon is preordained – in a similar way to the pinball machine of the last chapter where, even though some time must elapse before one valley is chosen over the other, initial conditions precisely determine the outcome. However, no theory has been constructed which (i) gives the same results for probabilities as quantum mechanics, and (ii) allows one to prepare the initial state of a photon so that it will activate one rather than another region of a photographic plate if quantum mechanics says that both are equally likely. Furthermore, no instance has been found in which the probabilistic predictions of quantum mechanics are falsified by experiment, whereas those of some classes of more detailed theories have been shown to be violated.†

In the 1930s an apparent resolution was provided by Niels Bohr (1885–1962), the so-called 'Copenhagen Interpretation,' which says that photons should not even be thought of as having definite properties independent of the apparatus that measures them. This interpretation attributes the loss of information (associated with the squaring of amplitudes and the occurrence of probabilities) to the process of measurement. In the present case the apparatus is the film. But, quantum mechanics claims to describe the interaction of a photon and a film, so there is no point at which some bridge from amplitudes to probabilities is crossed ...

I have written an endless loop into the last two paragraphs because if your head is like mine it starts right about now to spin. Usually, like most physicists with work to do, I simply square amplitudes and get on with it, accepting the fact that the lack of determinism in quantum mechanics is fundamentally different from anything previously encountered in physics. Einstein, among others, believed that quantum mechanics must consequently be incomplete. However, to restate a previous however, no one has found some deeper level of description which is capable of yielding new and correct predictions. And that is where the matter stands: a way of thinking especially suited to the gambling salon again seems to be essential if not for understanding then at least for predicting the inner workings of nature.

Quantum statistics

It would be a great mistake to end this chapter leaving even the slightest impression that there is something vague about quantum mechanics. This impression may also be left on the unwary by the words Uncertainty Principle, the usual rendering into English of Heisenberg's 1927 *Unbestimmtheitsprinzip*, which is in fact a precise statement of the limitations inherent in quantum mechanics on the simultaneous applicability of the classical concepts position and velocity, and is not

---

† I am referring here to theories satisfying 'Bell's Inequality,' to which references will be found in the suggested further readings.

uncertain at all. [See solved problem (2).] Users of quantum theory
never have any doubt about the calculational rules, and the mere fact
of probabilistic predictions will not trouble anyone who understands
that chance has its own powerful mathematics (Chapters 2–5). Although there seems to be as yet no answer to the question of why the
rules work, it is important to emphasize that they work exceedingly
well. They are the basis for the most precise calculations ever done.†

The quantum era began in the context of Statistical Mechanics.
Planck introduced the concept to explain experimental measurements
of the thermal properties of radiation. Since we have spent some
time thinking about heat, it seems appropriate to sketch these early
developments to illustrate quantum mechanics in use. The discussion
does not follow the historical order of events.

In the last chapter we re-encountered the 'harmonic oscillator,' a
mass connected to a spring and free to move in one direction, whose
energy is written in (11.1) and illustrated in Fig. 11.2. This, on the
face of it rather artificial, mechanical system was discussed in solved
problem 4 at the end of Chapter 6, where its period (or reciprocal
frequency) of oscillation was calculated. It is worth stating why this
is a useful model to consider. Many complicated mechanical systems
are found to be mathematically equivalent to a set of almost independent, i.e. weakly interacting, harmonic oscillators. Furthermore, an
electromagnetic vibration of a given frequency can also be thought
of as a harmonic oscillator, which trades energy not between kinetic
and potential energy but between electric and magnetic energy. In the
early days of quantum mechanics there was confusion about whether
a fundamental distinction is or is not to be made between oscillations
of mechanical systems on the one hand and of radiation on the other.
Einstein early came down in favor of the correct answer: the two can
be treated in the same way in quantum mechanics, just as they can be
in classical mechanics.

When modern quantum mechanics was created at the hands of
Heisenberg in 1925, the first problem he solved was the harmonic
oscillator. He found that an oscillator has states of motion labeled by
a non-negative integer, call it n, and given by the formula

$$\epsilon_n = (n + \tfrac{1}{2})h\nu, \qquad n = 0, 1, 2\ldots \tag{12.7}$$

Above, ϵ_n are what in quantum mechanics are called the energy
levels, ν is the frequency of the oscillator, and h is Planck's constant
[encountered in (12.3)] which is introduced into the theory precisely

† The magnetic moment of the electron has been calculated using these ideas to better than one part
in 10^{11}, by T. Kinoshita and his collaborators, and agrees with experiment to this accuracy.

to produce a spacing of levels in (12.6) by hv.† Although Planck and Einstein had guessed (12.7) – without the $\frac{1}{2}$, – the new theory gives insight into why only discrete energies are allowed. It may help to recall situations where discreteness is the rule: a violin string of fixed length and tension can only be made to vibrate at multiples of its lowest frequency; an old fashioned horn without valves has a similar property. Indeed, in the quantum theory a particle-like object, a point mass, has a 'waviness' associated with it; when the particle is trapped by a potential the wave-like behavior is controlled as are the vibrations of the string on a fiddle.

Now, discrete energy levels are very easily accommodated into statistical mechanics as presented in Chapter 8. We can contemplate the 'probability' (in the sense of thermal physics) of a particular quantum state in equilibrium. The Gibbs distribution for a collection of weakly interacting oscillators each with frequency v follows immediately from (12.7), and is

$$p_n = \frac{1}{Z} e^{-\beta \epsilon_n} \tag{12.8}$$

where

$$Z = \sum_{n=0}^{\infty} e^{-\beta \epsilon_n} \tag{12.9}$$

Furthermore, (12.8) and (12.9) then allow one to calculate the average energy per oscillator as follows:

$$<E> = \sum_{n=0}^{\infty} \epsilon_n p_n \tag{12.10}$$

The sums indicated above are done in solved problem (3) at the end of this chapter. The answer is

$$<E> = \left[\frac{1}{e^{\frac{hv}{T}} - 1} + \frac{1}{2}\right] hv = \left[N(\frac{hv}{T}) + \frac{1}{2}\right] hv \tag{12.11}$$

Here, T, the absolute temperature in energy units, is as always equal to the reciprocal of β, and the last equality defines the so-called Planck distribution. Comparing (12.11) with (12.7) one sees that the Planck distribution is the thermal average of the 'occupation number' n of the oscillator.

The plot in Fig. 12.3 is of the average energy as given in (12.11), minus the constant $\frac{1}{2}hv$. Two limiting forms of (12.11) are easy to

---

† In some formulations of classical mechanics a quantity with the same dimensions as Planck's constant is called the 'Action.' Thus h is often called the quantum of action.

obtain. When the temperature is high, in the sense $T >> h\nu$ the exponential in the denominator may be expanded according to (5.10), since the quantity being exponentiated is then very small. In the leading term that survives, the $h\nu$ cancels out giving $< E >= T + \dots$ However, if the temperature is comparable to or lower than $h\nu$ the average energy declines, ultimately becoming the exponentially small quantity $< E >= h\nu e^{-(h\nu/T)}$.†

The high temperature behavior in Fig. 12.3 is labeled 'classical limit,' because $< E >= T$ also follows from using the classical Gibbs distribution for energy states given by (11.1). The classical result is closely related to the expression for the average energy of an ideal gas molecule in equilibrium, $< E >= \frac{3}{2}T$, which we worked out in Chapter 8. Here one has kinetic energy (always quadratic in the velocity) and potential energy (in this particular case quadratic in the displacement.) Each of these, however, here is for motion in one (and not three) directions and as a consequence each contributes one

Fig. 12.3. Average energy of a quantum harmonic oscillator as function of Temperature.

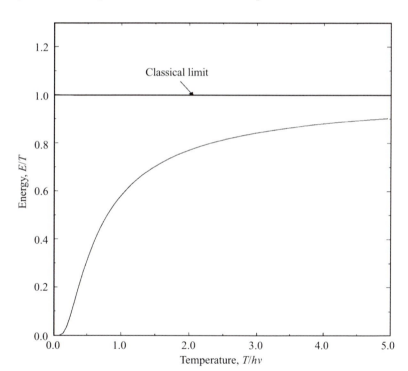

† It is a curious but true fact that the quantum era commenced in October 1900 with Planck guessing the formula (12.11) as the simplest way of interpolating between these high and low temperature limits, which were known to him from experimental sources. Within a few months he derived the formula in a way which, looking back on it, is quite correct, but must have seemed very strange indeed at the time. [See solved problem (5).]

third of the single molecule kinetic energy, namely $\frac{1}{2}T$, to the average energy, leading to a total classical average energy of T.

This explanation of why the average thermal energy of N harmonic oscillators is in general less than NT was one of the early triumphs of quantum mechanics. Planck was concerned with the average thermal energy of electromagnetic radiation in an enclosure. A classical description of this problem runs into serious trouble because according to Maxwell's theory, and the most modern ideas, vibrations at all wavelengths (no matter how small) – and thus through (12.1) at all frequencies (no matter how large) – are possible. In short, an infinite number of electromagnetic vibrations are permitted in any volume of empty space. Since these vibrations are in every respect harmonic oscillators, each of them according to classical ideas must have an average energy of T in equilibrium. An infinite amount of energy would then be needed to increase the temperature of a finite volume of empty space! Planck's formula (12.11) saves the day by showing that vibrations of high frequencies are 'frozen out,' as Fig. 12.3 indicates. Without entering into the details, perhaps it will suffice to state that when the average energy (12.11) is associated with each 'radiation oscillator,' the thermal properties of radiation are completely and quantitatively explained.

The calculation presented here was first done by (again!) Einstein, in 1906, and suggested by him as the explanation for why the average energy of solids in equilibrium at low temperatures is found by thermal measurements to be less than the classically expected amount. The details of his explanation are wrong because he made too crude a model for the (finite) set of harmonic oscillators that describe the collective vibrations of a solid, but the essential physical idea is dead right and preserved in any text on solid state physics.

Emission and absorption of light

Without conscious intention on my part, this chapter has turned into a survey of Einstein's early heroic and revolutionary contributions to quantum mechanics. I shall close with a sketch from a more modern perspective of his 1916 discussion of the interaction of light and matter. This very early work allows one to understand much of the essential physics underlying the operation of a the modern quantum optical device, the laser.†

---

† An acronym for L(ight) A(mplification) by S(timulated) E(mission) of R(adiation).

As background it is necessary to know that in quantum mechanics energy is conserved during the emission and absorption of light. This can be expressed in a form first given by Niels Bohr (1885–1962),

$$E_2 - E_1 = h\nu. \tag{12.12}$$

Here E_2 and E_1 ($E_2 > E_1$) are two discrete energy states of a group of atoms able to emit and absorb photons, which we shall henceforth call an optically active 'center.' Imagine a finite volume of transparent matter which can sustain an electromagnetic vibration at frequency ν, and containing centers of this kind. The vibrations behave precisely like a harmonic oscillator as considered in the last section, and in equilibrium at temperature T will have a mean energy as given in (12.11).

A useful identity obeyed by the Planck distribution $N(\nu, T)$ defined in (12.11), which you should please check is

$$\frac{N(\nu, T) + 1}{N(\nu, T)} = e^{(h\nu/T)} \tag{12.13}$$

In equilibrium at the same temperature T, the energy states E_2 and E_1 will have thermal probabilities given by a Gibbs distribution as in (12.8). If p_2 and p_1 are the respective probabilities we have

$$\frac{p_1}{p_2} = e^{(E_2 - E_1)/T} = e^{(h\nu/T)}, \tag{12.14}$$

where in the last line we have used (12.12). Now using (12.13) we find that in equilibrium between light and a 'center' the following condition holds,

$$p_1(T)N(\nu, T) = p_2(T)[N(\nu, T) + 1]. \tag{12.15}$$

An intuitive interpretation can be given of (12.15). The left hand side, the product of the probability for the center to be in its lower state and the thermal occupation number of the radiation oscillator, is proportional to the average rate at which (in equilibrium) state 1 is excited to state 2. Since in equilibrium the average excitation rate from state 1 to state 2 must be equal to the average de-excitation rate from 2 to 1, the latter must be proportional to the right hand side.

The expressions on the left and right of (12.15) are indeed proportional to the excitation and de-excitation rates in the modern theory of quantum optics. These processes deplete and repopulate the lower state, and the net rate of reduction of the probability of state 1 is found to obey the 'rate equation'

$$\frac{\Delta p_1}{\Delta t} = -\Gamma[p_1 N - p_2(N + 1)] \tag{12.16}$$

where Γ, the 'decay probability,' is calculable from quantum mechanics. Inside the bracket on the right is the difference of the left and right hand sides of (12.15), which difference is not zero away from equilibrium where N is not the thermal average of n but the actual state of excitation of the radiation oscillator, which we now call the 'number of photons,' and p_1, p_2 also do not necessarily correspond to equilibrium. In a subtle heuristic argument showing how the Planck distribution emerges from a consideration of equilibrium as a steady-state of balance between emission and absorption, Einstein in 1916 intuited an equation equivalent to (12.16), more than a decade before the full theory was invented! Evidently, the assumption of a steady state, $(\Delta p_1/\Delta t) = 0$, in (12.16) will give (12.15), and the assumption of a Gibbsian equilibrium ratio (12.14) between p_2 and p_1 will then yield (12.13) whose solution is Planck's function.

We can put (12.16) to two uses. First, as promised, we can deduce the essential physical principles of the laser. Second, we can bring into focus Einstein's first worries about quantum mechanics.

Lasers work because when $N \gg 1$, and p_2 is greater than p_1, the left hand side of (12.16) is large and positive, corresponding to a large net rate of $2 \rightarrow 1$ transitions. Since each de-excitation produces another photon, this process (called stimulated emission) can be used to amplify light – explaining the acronym laser. The process is central to, for example, maintaining the signal in fiber optic technology. One requires a non-optical way of producing a larger population of centers in the 'excited state' 2 than in the 'ground state' 1. This is typically done by choosing a center that has a third state 3 with energy E_3 greater than E_2, and populating 3 using a steady light source at the 'pump' frequency $v_p = (E_3 - E_1)/h$. The state 3 is chosen to have the property that it can decay to state 2 giving up the energy $E_3 - E_2$ to the optical medium without emitting light. A 'population inversion' [the negative temperature referred to on p. 137] is thus induced, which now amplifies light at the laser frequency $(E_2 - E_1)/h$. This is the basic theory of the laser, a triumph of quantum mechanical reasoning.

As always with quantum mechanics, something odd is lurking about in (12.16). The 1 in the second term on the right hand side of (12.16) – which 1 we have seen to be essential for the Gibbs and Planck distributions to be mutually consistent – means that de-excitation of the center can happen spontaneously, i.e. with no quantum of light present.† The embarrassing question is, When does such a decay happen? Quantum mechanics only ascribes a probability per unit time

---

† This is nothing more or less than an example of what we called radioactive decay in Chapter 5.

for the process to occur, and contains no hidden gears or wheels to predict (even in principle) exactly when a particular jump will occur. In short, this *when?* is answered with intrinsic probability, just as is the *where?* of photography. Einstein did not like this answer, and he never reconciled himself to it.

Quantum mechanics has become an applied science – a useful tool for engineering, dealing perfectly, as far as we know, with all phenomena. Is it silly to ponder its conceptual foundations? No one can be sure. There are those who think that something remains to be found here which will yield a yet better understanding of nature. Let Feynman have the last word: 'We have always had a great deal of difficulty in understanding the world view that quantum mechanics represents... It has not yet become obvious to me that there's no real problem. I cannot define the real problem, therefore I suspect there's no real problem, but I'm not sure there's no real problem.'

Further reading

My source for information about Einstein has been the biography by Abraham Pais, *Subtle is the Lord*, (Oxford 1982). His biography of Bohr, *Niels Bohr's Times* (Oxford 1991), is another excellent source for the early history of quantum mechanics. These books probably cannot be read in detail without a systematic education in physics, but you may be pleasantly surprised at how much you will be able to get out of them. Feynman explains quantum mechanics to a general audience in his *QED: The strange theory of light and matter* (Princeton 1985). There is an accessible and deep discussion of the foundations of quantum mechanics in Anthony J. Leggett's *The Problems of Physics* (Oxford 1987). Section II of David Mermin's *Boojums all the way Through* (Cambridge 1990) contains several thought-provoking articles about quantum mechanics. The last two references contain discussions of 'Bell's Inequality.' A book intended for teaching non-scientists about light (and the special theory of relativity) is Ralph Baierlein *Newton to Einstein* (Cambridge 1992). In discussing the early history of light, I simply reinterpret the essential ideas in a modern context. To have any notion of the mental processes of the time, one must read original sources. A recent scholarly book about some of the matters touched on here is Jed Z. Buchwald *The Rise of the Wave Theory of Light* (Chicago 1989.)

Solved problems

(1) Calculate how many photons per second are radiated by a light bulb which emits 10 watts of bluish-green light of wave length 0.5 μm. One watt of 'power' means an energy emission rate of 1 joule per second.

Solution: We must first calculate the quantum of energy for light of the given wave length via $h\nu = hc/\lambda$. Taking $h \approx 6.6 \times 10^{-34}$ J s (joule second) and $c \approx 3 \times 10^8$ m/s (meters per second) one finds $h\nu \approx 4 \times 10^{-19}$ J. Now, we are told that 1 watt is 1 J/s. Let P be the power emitted, 10 W. Then the number of photons emitted per second is $P/h\nu$, which comes out to be an enormous number of approximately 2.5×10^{19} photons per second.

(2) For a particle of mass M, the Heisenberg Uncertainty Principle is the inequality

$$M\Delta x \Delta v \geq \frac{h}{4\pi} \tag{12.17}$$

where Δx and Δv refer to the precision with which the classical notions of position and velocity may be used simultaneously in quantum mechanics. The restriction is a consequence of the properties of 'wavicles.' The product $M\nu$, called the momentum, is inversely proportional to the wavelength of the amplitude we called Ψ. To localize a wavicle in quantum mechanics requires putting together many wavelengths. A well localized wavicle has to have a spread of wavelengths and thus velocities.

In this solved problem, we are going to use (12.17) to show that the lowest possible energy of a harmonic oscillator is $\frac{1}{2}h\nu$ as in (12.7). The total energy of a harmonic oscillator is

$$E = \tfrac{1}{2}M\nu^2 + \tfrac{1}{2}kx^2. \tag{11.1}$$

In classical mechanics, the lowest energy corresponds to the state of rest $\nu = 0$ at $x = 0$. This violates (12.17) and is not possible in quantum mechanics. The best one can do is to relate velocity and position via the smallest product consistent with (12.17),

$$\nu = \frac{h}{4\pi Mx}. \tag{12.18}$$

Substitute this condition in (12.17) and find the lowest possible energy.

Solution: Making the suggested substitution (12.18) in (11.1), we get

$$E = \tfrac{1}{2}\left(\tfrac{h}{4\pi}\right)^2 \tfrac{1}{Mx^2} + \tfrac{1}{2}kx^2. \tag{12.19}$$

Note that at very small x the first term (kinetic energy) becomes large because of the x^2 in the denominator. The best compromise is achieved by working out how E changes when x is changed by a little. It is just as well that we used the symbol x in (12.19) for what is in fact the spread in x, because we now want to contemplate small changes in this spread, for which we have always used the Δ notation. From results worked out in chapter 5, one has

$$\Delta\left(\frac{1}{x^2}\right) = \frac{1}{(x + \Delta x)^2} - \frac{1}{x^2} \approx -2\frac{\Delta x}{x^3} \tag{12.20}$$

$$\Delta(x^2) = 2(\Delta x)x,$$

from which it follows that the flat point of E, a minimum as one sees by sketching (12.19), occurs for $x = x_m$, where

$$kx_m = \left(\frac{h}{4\pi}\right)^2 \frac{1}{Mx_m^3}.$$ (12.21)

Putting this best compromise into (12.19) and recalling from problem (4) of Chapter 6 that the frequency of classical oscillation is $v = (1/2\pi)\sqrt{k/M}$, one finds the minimum energy to be the advertised $\frac{1}{2}hv$.

(3) Calculate the mean energy in thermal equilibrium at temperature T of a quantum harmonic oscillator with frequency v. You will need to do sums of the form $(a < 1)$

$$1 + a + a^2 + a^3 + \cdots = \frac{1}{1-a},$$ (12.22)

which you will find in solved problem (2) in Chapter 5, and

$$a + 2a^2 + 3a^3 + 4a^4 + \cdots = \frac{a}{(1-a)^2}.$$ (12.23)

The latter can be proved from the former by replacing a by $a + \lambda$ and working out a multiplied by the coefficient linear in λ on both sides, using solved problem (1) at the end of Chapter 5. It also follows from squaring the left and right of (12.22) and multiplying both sides by a.

Solution: This is an application of the Gibbs distribution. One has to work out (12.10), using (12.7–9).

The denominator Z in the probability distribution (12.8) is

$$Z = e^{-\frac{hv}{2T}}[1 + a + a^2 + \cdots]$$ (12.24)

where

$$a = e^{-\frac{hv}{T}}.$$ (12.25)

Summing the series gives

$$Z = e^{-\frac{hv}{2T}} \frac{1}{1 - e^{-\frac{hv}{T}}}.$$ (12.26)

Now, substituting (12.7) into (12.10) gives

$$<E> = \frac{e^{-\frac{hv}{2T}}}{Z} \sum_{n=0}^{\infty} (n + \frac{1}{2})hv\, e^{-n\frac{hv}{T}}.$$ (12.27)

Note that $\frac{1}{2}hv$ is a common factor to a sum of the form already done, and the remaining sum is like the second one given in the statement of the problem. When this is carried out, the result is

$$<E> = hv\left[\frac{e^{-\frac{hv}{T}}}{1 - e^{-\frac{hv}{T}}} + \frac{1}{2}\right].$$ (12.28)

The first ratio in the square bracket can be simplified by multiplying numerator and denominator by $e^{\frac{hv}{T}}$, whereupon the result is (12.11).

(4) Calculate the entropy S of a quantum oscillator in equilibrium at temperature T, using the method of chapter 8, i.e.

$$S = -\sum_n p_n \ln p_n \qquad (8.11)$$

Solution: Using (12.7) we have $\ln p_n = -(\epsilon_n/T) - \ln Z$. Substituting this into (8.11) and using (i) $\sum_n (p_n \ln Z) = \ln Z$; (ii) the expression (12.26) for Z that was calculated in the last problem; and (iii) the fact that $\sum_n \epsilon_n p_n =< E >$, given in (12.11) and calculated in the last problem, we get

$$
\begin{aligned}
S &= \frac{1}{T} < E > + \ln Z \\
&= \frac{1}{T} hv \left[N(v, T) + \tfrac{1}{2} \right] - \frac{hv}{2T} - \ln[1 - e^{-\frac{hv}{T}}] \qquad (12.29) \\
&= \frac{hv}{T} N(hv, T) + \ln[N(v, T) + 1]
\end{aligned}
$$

a neater form is obtained using the identity obtained by taking the logarithm of both sides of (12.13), whereupon it follows that

$$S = [N(v, T) + 1] \ln[N(v, T) + 1] - N(v, T) \ln N(v, T) \qquad (12.30)$$

(5) Derive the formula

$$S = (N + 1) \ln(N + 1) - N \ln N \qquad (12.31)$$

for the entropy of a quantum oscillator in the following way. Count the number of ways C of distributing Q identical quanta of excitation among D oscillators. Evaluate $\ln C$ for large D and Q using Stirling's formula and call Q/D, the mean number of quanta per oscillator, N. Show that $\ln C = D \times S$ where S is given by (12.31).

Solution: Let us do the counting step by step. Consider D long boxes side by side, in each of which the state of excitation is given by a row of marbles, the number of them denoting the number of quanta of excitation in that particular oscillator. We have Q marbles to distribute among the D boxes. First consider numbered, and thus distinguishable, marbles. Marble 1 can be placed in any one of the boxes, and thus in D ways. For each one of these D ways, Marble 2 can be placed in $D+1$ ways – because it can be put in any one of the $D-1$ empty boxes and to the right or left of the marble 1 in whichever box it is in. By similar reasoning, marble 3 can be placed in $D+2$ ways, for every arrangement of 1 and 2. At every stage the number of ways of adding an additional marble to a given box is one more than the number already in that box. For all D boxes the number of ways of adding an additional marble is thus D plus the total number of marbles already placed. In short, the number of ways of distributing Q distinguishable marbles is $D(D+1)(D+2)(D+3)...(D+Q-1)$, a total of Q factors. To correct for the indistinguishability of the quanta, we have to divide by $Q!$, the number of rearrangements of the Q marbles. We thus get

$$C = \frac{[D(D+1)(D+2)...(D+Q-1)]}{Q!} = \frac{(D+Q-1)!}{(D-1)! \, Q!}. \qquad (12.32)$$

Using Stirling's formula for large N, namely $\ln N! \sim N \ln N - N$, for each of the factors in the last expression in (12.32), one obtains

$$\ln C = D\{(1 + N - \frac{1}{D}) \ln[D(1 + N - \frac{1}{D})]$$

$$- (1 - \frac{1}{D}) \ln[D(1 - \frac{1}{D})] - N \ln(DN)\}, \tag{12.33}$$

where we have replaced Q/D by N. Now, we may neglect all terms $1/D$ because $D \gg 1$ by hypothesis. Displaying the terms proportional to $\ln D$ using $\ln DX = \ln D + \ln X$, for any X, one sees that the sum of such terms is zero. The quoted result is then obtained.

(6) *Assume* the following equations for the entropy S and the energy E of a quantum harmonic oscillator

$$S = (N + 1) \ln(N + 1) - N \ln N,$$

$$\text{and} \quad E = h\nu N \tag{12.34}$$

where N, the mean number of quanta per oscillator is unknown. Find N by requiring that for small changes $\Delta E = T\Delta S$, where T is the temperature. This condition imposes the first [(9.8)] and second [(9.14)] laws of thermodynamics when no work is done (i.e. $\Delta W = 0$), and is analogous to the condition used to derive the Gibb's distribution in chapter 8.

Solution: This is Planck's 1901 derivation of his distribution. (He obtained the expression for S from counting states in a manner similar to that in solved problem (5).)

Contemplate a small deviation ΔN from the sought function. Then, since [(8.11)] $\Delta[x \ln x] = (\ln x + 1)\Delta x$, we get

$$\Delta S = \Delta N[\ln(N + 1)] + \Delta N - \Delta N[\ln N] - \Delta N$$

$$= \Delta N \ln[\frac{N + 1}{N}], \tag{12.35}$$

$$\text{and} \quad \Delta E = h\nu\Delta N.$$

The condition to be imposed is then

$$\Delta N[h\nu - T \ln \frac{N + 1}{N}] = 0, \tag{12.36}$$

which can only be satisfied in general if the square bracket is zero. You will verify that this condition produces the Planck distribution.

Since the Planck distribution

$$N\left(\frac{h\nu}{T}\right) = \frac{1}{e^{\frac{h\nu}{T}} - 1} \tag{12.37}$$

has become such an important part of this last chapter, it is worth mentioning that its most precise experimental verification came in 1990 from the COBE (cosmic background explorer) satellite, which was designed to detect the residual photons expected from the 'big bang' theory of the origin of the universe and measured a thermal spectrum of radiation characterized by the temperature 2.726 ± 0.005K, filling space. At the start of this book, I wrote that in physics there is an unusual kinship through the ages. The calculation described in the last problem, which resolved a major experimental challenge to the classical physics of the time, is almost one century old. In time, it may have to be modified to deal with phenomena yet unknown. But in its domain of applicability – which, astonishingly, includes outer space – there is nothing antique or quaint about it.

Index